Drinking Water: A Socio-economic Analysis of Historical and Societal Variation

In this fascinating and challenging work, the author analyses the way water for drinking is produced, distributed, owned, acquired and consumed in contrasting ways in different settings. From the taken-for-granted, all-purpose water, flowing out of taps in advanced economies to extreme inequalities of access to water of variable qualities, drinking water tells its own interesting story, but also reflects some of the centrally important characteristics of the state and economies of different countries. From sparkling mineral water in Germany, to drinking water garages in Taiwan, from water tankers in Mexico City to street vendors in Delhi markets, comparisons are made to stretch our understanding of what we mean by 'an economy', quality, and property rights of water.

In addition, the study of the socio-economics of drinking water provides a route into understanding interactions between polity, economy and nature. One of the major themes of the book is to analyse the 'sociogenic' nature of sustainability crises of economies of water in their environmental settings: epidemics, droughts, pollution, land subsidences and floods. Overall it develops an economic sociology, neo-Polanyian approach in a comparative and historical exploration of water for domestic consumption.

Mark Harvey is a Professor in the Department of Sociology and Director of the Centre for Research in Economic Sociology and Innovation, University of Essex, UK.

"Theoretically generative, methodologically creative, richly comparative, and politically engaged, *Drinking Water* doesn't just work within the Polanyian tradition, it substantively reinvents that tradition. Working imaginatively around the nexus of political economy and political ecology, Mark Harvey and his colleagues confront the question of urban drinking water not just as a 'policy problem,' but as a departure point for a bold exploration of the diverse *properties* of water – as an unruly public good, as an uncooperative commodity, as a mercurial space of state regulation, as an enduring biosocial necessity, and more. This is a truly original and important book."

– Jamie Peck, University of British Columbia, Canada

"A fine and fascinating book, exploring the inter-relationship of natural resources, property rights, economic organisation, forms of exchange and modes of consumption, and offering intriguing detail, telling analysis, theoretical sophistication and political relevance."

– Alan Warde, University of Manchester, UK

"A major contribution to political economy that explains how drinking water came to flow from tap in some countries but is purchased in bottles in others."

– Frank Trentmann, Birkbeck, University of London, UK

Drinking Water: A Socio-economic Analysis of Historical and Societal Variation

Mark Harvey

With contributions from Adrian Evans,
Daniel Figueroa, William D.H. Li
and Aviram Sharma

Routledge
Taylor & Francis Group
LONDON AND NEW YORK

First published 2015
by Routledge
2 Park Square, Milton Park, Abingdon, Oxon OX14 4RN

and by Routledge
711 Third Avenue, New York, NY 10017

First issued in paperback 2017

Routledge is an imprint of the Taylor & Francis Group, an informa business

British Library Cataloguing-in-Publication Data
A catalogue record for this book is available from the British Library

Library of Congress Cataloging in Publication Data
Harvey, Mark, 1943-
Drinking water : a socio-economic analysis of historical and societal variation / Mark Harvey ; with contributions from Adrian Evans, Daniel Figueroa, William D.H. Li, and Aviram Sharma.
pages cm
Includes bibliographical references and index.
Drinking water. 2. Water-supply. 3. Drinking customs. 4. Beverage industry. I. Title.
TD220.2.H37 2016
363.6′1--dc23
2015016081

ISBN 13: 978-1-138-30497-0 (pbk)
ISBN 13: 978-1-138-81697-8 (hbk)

Typeset in Times
by Saxon Graphics Ltd, Derby

Contents

Acknowledgements

The research undertaken for this book was funded by the Economic and Social Science Research Council, the British Academy and the National Science Council of Taiwan. The ESRC research was part of a programme of the Sustainable Practices Research Group (Res-597-25-0003). I would particularly like to thank the Group Directors, Professors Dale Southerton and Andrew McMeekin, as well as Professor Alan Warde, for all their support and encouragement. The Taiwanese component of the research was funded by the British Academy and the Taiwan National Science Council under an International Partnership and Mobility award.

Authors

Professor Mark Harvey is Director of the Centre for Research in Economic Sociology and Innovation (CRESI), in the Department of Sociology at the University of Essex. His recent books include *Exploring the Tomato: transformations of nature, economy and society*; *Trust in Food: a comparative institutional analysis*; *Public or Private Economies of Knowledge: Turbulence in the Biological Sciences*; *Karl Polanyi: new perspectives on the place of the economy in society*; and *Markets, Rules, and Institutions of Exchange.*

Dr Adrian Evans is a senior research fellow at The Centre for Agro-ecology Water and Resilience (CAWR) at Coventry University. He was previously a researcher for the ESRC Sustainable Practices Research Group (SPRG) at the University of Essex from 2011 to 2013, developing the IEP (Instituted Economic Processes) approach to understand the formation of the markets for bottled water in the UK, Italy and Germany. His main research interests include: consumption, theories of practice and embodiment, food ethics, food consumption, farm animal welfare, religious slaughter and facilitating science–society dialogues around food and farming issues.

Dr Aviram Sharma gained his doctorate at the Centre for Studies in Science Policy, Jawaharlal Nehru University, New Delhi, India. His doctoral thesis deals with the regulatory decision making for bottled water quality in India. His research interest lies on the interface of science and technology studies, public understanding of science, environmental studies, economic sociology and heterodox economics.

Dr Daniel Figueroa completed his PhD on the configuration of socioeconomic processes of internet diffusion in Brazil and Mexico at the University of Essex in 2014. He is currently Director of the Market Research Department at Blank Consulting, located in Mexico City. His latest research and current interests include competence mapping and educational divide analysis in engineering positions in the automotive industry in Mexico. He is currently researching political affiliation and its relation with the perception of different socioeconomic problems and overall life satisfaction.

Professor William D.H. Li is Professor of Sociology at the National Dong Hwa University, Taiwan. His research has centered on social change and social welfare with special reference to East Asia. He is the author of *Housing in Taiwan: Agency and Structure?* (Ashgate, 1998), numerous journal articles and book chapters. Currently he is undertaking international comparative research on the East Asian welfare model, housing, living arrangements, and economic sociology on drinking water.

Figures

Tables

1 Properties of water

Dipping into water

Properties and properties

Properties of water can mean many things: characteristics of water, physical, cultural, natural, or social, for example. Drinking water or spiritually cleansing water. Mountain spring water, or deep ocean water 'untouched by human civilisation', as one Taiwanese company bottling it bizarrely claimed. Swimming pool water or rice paddy field water. The water in Coca Cola or the water flushed down toilets. Tea or soup. Land or sea water.

But 'property' can also refer to ownership: what kind of ownership and by whom? If the public owns it, what public? If the state, what water (rivers or groundwater aquifers) and on what scale within what territorial or cross-territorial boundaries? If a local community, by virtue of their 'traditional' collective proximity to a water resource? Moreover, is a proprietorial right legal or illegal, formal or informal, and, perhaps at least as significantly, are there proprietorial water voids: ambiguities, uncertainties or absences of rights to control a water resource?

And then there are many ways of enquiring or knowing about water properties of all kinds: folk knowledge or natural scientific hydrology, market research or environmental science, anthropology or chemistry, urban planning or agronomy.

Dipping into water, this book is an exploration of a particular kind into a particular kind of water: water for domestic consumption, primarily drinking. Initially, the springboard for the enquiry was the controversy surrounding the sustainability of bottled water, a rapidly expanding consumption habit and market in the United Kingdom, in relation to tapwater (Barlow and Clarke, 2002; Swiss Gas and Water Association, 2006; Botto, 2009; Gleick and Cooley, 2009; Gleick, 2010). In a modern developed urban society, the taken-for-granted ready availability of drinkable water flowing out of taps is only quite exceptionally undermined by droughts or burst mains. Yet, the very ordinariness of tapwater first needs to be seen in the perspective of how this urban tapwater went from extraordinary to ordinary – a process that took over a century and a half in the case of the United Kingdom. Second, its characteristic property of all-purpose water, for drinking, cooking, washing bodies and clothes, flushing toilets, gardening, again an historically manufactured property, must equally be seen for its very peculiarity.

Tapwater has a doubly extraordinary ordinariness, always on tap, used for all domestic purposes. Not dedicated drinking water. That, you find in bottles.

Moreover, as soon as you leave the comfort zone of modern urban infrastructures, to places such as Delhi or Mexico City where both these ordinarinesses are entirely absent, you are confronted with a variety of water types, many of dubious quality, where even basic access to water of any kind can become a challenge. In such places, and for those that can afford it, bottled water may be the only readily available drinkable water. So, Mexico City has the highest per capita consumption of bottled water in the world, supplemented by very high levels of soft drink consumption. Apart from going back in history, therefore, comparing London with Delhi reveals the peculiarity of our ordinariness with respect to water. As will be recounted later, our exploratory dipping into water will take us to three capital cities (London, Delhi, Mexico City), to Taiwan, and to a comparison of the provisioning and consumption of bottled water in the United Kingdom, Germany and Italy.

Water and the wider economy

So far, we have been discussing drinking water primarily for its own characteristic properties, treating water as the *object* of understanding. Switching attention to ownership and control, properties of water assume an economic dimension. An essential good for human survival, water has had many turbulent political and economic histories (Sultana and Loftus, 2011), its scaled-up channelling and control, for example, being linked to the historical emergence of centralised state power, even despotic regimes, by Wittfogel (1963) and Marx (Godelier, 1970; Molle, Mollinga and Wester, 2009; Bakker, 2002). Contemporary China is witness to the world's largest ever hydraulic engineering construction providing water to Beijing, the South–North Water Transfer Project, a channelling of water enabled by the exercise of state power of a particular kind (Wang and Huang, 2013; Rong, 2011). One strand of political ecology emphasises the linkages between flows of water, flows of power, and, with the 'commodification' of water turning it into a good with a monetary value, flows of money (Swyngedouw, 2004; Kaika and Swyngedouw, 2000; Heynon, Kaika, and Swyngedouw, 2005). Water, and especially tapwater for domestic consumption, however, has been characterised as an 'uncooperative commodity' (Bakker, 2003a) because of its frequently incomplete marketisation and conflicts over quality and price regulation. Water provision, always highly politicised, has been beset with turbulence around its privatisation, marked by 'governance failures' in many parts of the world (Bakker, 2010). Conversely, water can also be seen as an 'unruly' public good on two main counts. Firstly, the scope and scale of the 'public' for water as a resource (local, regional, national, international) is often contested and unstable, including whether a modern nation state 'owns' its territorial water resources, and in what sense. It may come as a surprise that water resources in the United Kingdom were only legally appropriated as kind of national-scale property in the 1980s (Hassan, 1998; Bakker, 2003a). Secondly, the nature and extent of a public's obligation and capacity to provide water and of what quality and accessibility, continues to be a matter of contestation in many parts of the world, even in terms of meeting the Millennium Development Goals (Gleick, 2011).

Exploring water as an economic good, therefore, provides a reflecting surface for understanding the broader political economy of a given society: how or whether a society establishes universal rights to water to meet a fundamental need, for example, and the extent to which this is related to the development of political democratic rights. The right to vote and the citizen's right to access to water were often historically interwoven. But there are paradoxes too: the USA as a heartland of free market capitalism retains water provision in the public domain in many of its major cities, including New York and Washington DC. Conversely, France, otherwise exemplary of the European social model, has relied on private capital to develop its metropolitan and urban water supplies throughout its history (Goubert, 1986; Millward, 2000, 2005, 2007). So water reflects significant variations in the often shifting place of the market and state in the economic organisation of a society.

Water: sustainability and nature

Water also represents an exemplary flow of interconnection between polity, economy and nature. The expansion and then collapse of the Mayan civilisation, with its water economy built around complex and highly developed infrastructures, has been attributed to successive periods of unusually plentiful rainfall succeeded by prolonged droughts (Haug et al., 2003; Kennett et al., 2012; Demarest, 2004). Even with the narrower focus on provision of domestic water for large cities, the particular interdependencies between quite localised water resources (such as a large river) and specific dynamics of societal and demographic change, frequently generate distinctive sustainability crises peculiar to a London or Beijing, a Delhi or Mexico City. These emergent sustainability crises will be one of the themes of this book. Delving into water, therefore, provides a powerful optic for exploring the relation between economy and nature.

The natural availability of water can also be taken for granted in stable historical and societal circumstances – past sustainability crises recede in societal memory, and can almost be epistemologically suppressed in certain conceptual representations. Water can be understood as a natural common pool resource (along with biodiversity, the atmosphere, fish stocks, and many other examples). There is even the idea of this common pool as a 'state of nature', prior to the emergence of any property rights. This concept is then buttressed by an almost 'state of nature' human, blessed with 'free access' to that naturally given pool (Ostrom et al., 2002; Agrawal, 2003).[1] Comparative anthropological and historical political economy reflections on water, however, shatter this vision of natural common pools and free

1 'When no property rights define who can use a common-pool resource and how its uses are regulated, a common-pool resources is under an open-access regime.' (Ostrom et al., 2002, p.19)

(in the sense of free for all) access. There certainly are natural ‘pools’ (rivers, seas, groundwater, rain, etc.) but equally certainly there are no natural human commons, given that all forms of social interdependencies with those pools involve differential access to them, even at the basic level of the location of different societies in different water environments. Likewise, there is no natural state of the ‘human’, where all members of a society or collectivity have the same and equally free access to ‘natural pool’ water resources as a given human condition.

Problematising the commons – the unruly public good – is therefore also a way of problematising the flows of interconnection between economy and nature. Some perspectives veer to a social constructionist view, where scarcities and abundances of water are socially, discursively and politically constructed (e.g. Kaika, 2003). Others suggest that social science can construct its own conceptualisation of a naturally given materiality in parallel to, and independently of, natural or environmental science constructs (e.g. Bakker and Bridge, 2006). In the more extreme versions of this perspective, this material world (parallel to, or bypassing, natural scientific discourse) is populated with ‘actants’, the physical universe as described by social scientists – or philosophers (Callon, 1986; Latour, 2004). The position to be developed here assumes that there are naturally given ‘pools’; that natural sciences are best placed to understand natural, non-human processes and their dynamics (e.g. the hydrological cycle, El Niño phenomena, etc.); but that social science gets into the act by exploring the interdependencies and interactions between societies and naturally given ‘pools’ – in this case, water. In undertaking its explorations, social science reliance on natural science conceptualisations of the given natural world is itself an aspect of the interdependencies and multiple interactions between societies and nature. We can choose to work in complementary rather than parallel universes.

A cameo of water complexity

The flows of interconnection between the natural and the social *and* the complementarity of social and natural science are supremely well illustrated by Evans’ *Death in Hamburg,* an analysis of what he describes as the ‘man-made epidemics’, especially of cholera, which resulted in extreme loss of life in that city during the nineteenth century (Evans, 1987). He provides us with a cameo of an emergent sustainability crisis centred on water and the flows of interconnection between the natural, the social, the economic and the political. The cameo, presented with some detail here, usefully suggests an historical way of thinking about the multi-dimensional complexities of water. Although predominantly a mercantile city, Hamburg became progressively industrialised, and shared with current examples of massive and rapid urbanisation the slum conditions, the sewage and water contaminations, that created the conditions for concentrated epidemics. However, distinctively for Germany, let alone other countries in Europe, the relatively autonomous city of Hamburg was dominated by ‘free-market’, ‘nightwatchman state’ politics of its mercantile elite government. Even before the cholera epidemic of 1892, the Senate had refused to impose qualification requirements on the medical profession as an infringement of market freedoms, and likewise had failed to impose smallpox vaccination (discovered by Jenner in 1789) on the grounds that every citizen had the right to decide what they did to their own bodies. In taking this stance, Hamburg contrasted with Prussia, where mandatory smallpox vaccination was instituted in 1818 as a condition for many occupations, enrolment into schools and recruitment into the army. Unlike the rest of Germany, therefore, Hamburg had one of the most devastating epidemics of smallpox as late as 1871, with nearly 5000 deaths. The bio-human sustainability crisis of the cholera epidemic, however, was absolutely tied to the politics and economies of water provisioning in the city. Schematically, the key aspects of these dynamics involved a partial establishment of a tapwater system to much of the city, by a British engineer, Lindley. However, although WCs partly replaced cesspits and open sewage disposal into the cities fleets (canals), both the harbour area and much of the river Elbe were exposed to human and industrial pollution. The tapwater system was eventually fed by a reservoir and source upstream of the city. But there was a long battle between various groups, merchants, landlords, small shopkeepers and industrialists that resulted in the refusal to build slow-sand filters to the water provision, unlike in the Prussian, and immediately adjacent town of Altona. Filtration of tapwater had become widespread in Europe by the late nineteenth century. Hamburg resisted it, to the end. So much for the elements for our contemporary understanding of the sociogenesis of epidemics in the city.

But a further sociogenic element was undoubtedly the contemporary medical science of infectious disease, and the understanding of the biological and natural causes, that, irrespective of human intervention, resulted in human biological death. The influence of one of the major hygienists of continental Europe, Max von Pettenkofer, was particularly strong in Hamburg. Evans argues that there was a concordance between his miasmatist views on disease, which rejected the efficacy of quarantines, disinfection and sanitary regulation, and a nightwatchman state that resisted any form of state intervention, especially ones that might restrict the flow of trade. So when the epidemic did break out, there was extreme reluctance for the state to intervene, endorsed by the city's medical profession then in the thrall of Pettenkofer. By contrast, the Prussian state had by that time been converted to contagionism, to the extent that it had supported a contemporary medical race between France (Pasteur) and Germany to find the cholera bacillus bearing the disease infection. Robert Koch led a German team to Egypt where there was a cholera epidemic in an attempt to beat the French to it. But when both teams arrived, the epidemic had died down, and the French team went home. Koch, undeterred, went on to Calcutta, and eventually successfully isolated, identified and grew the comma bacillus in a culture. The hero returned, and transformed both medical science and state policy in Berlin.

Given these preconditions, political, economic, social, cultural *and bio-environmental*, the precipitating trigger for the cholera epidemic was the arrival of migrants from Russia travelling to the USA via the main conduit port of Hamburg, and bearing the cholera bacillus. Soon the Elbe was contaminated, and, unfiltered, the cholera bacillus thrived and then was efficiently distributed to every household in the city through the tapwater system. Within a couple of weeks during Autumn 1892, nearly 10,000 people had died of cholera in Hamburg. By contrast, Altona, drawing water from the same source of the Elbe, but with sand filtration, was virtually unaffected by cholera.

The social divisions of rights to water and sanitation were powerfully expressed in mortality figures: houses with WCs largely escaped, whereas tenement blocks with collective cesspits were mortality hotspots. The city belatedly exhorted the population to boil all water before consuming it, given that high water temperature was a lethal environment for this particular biological organism. The employers of domestic servants, most of the upper and middle class who could also afford to boil water, had the lowest mortality figures. Domestic servants, handling contaminated water as part of their job, had the highest mortality figures of all occupations, followed closely by harbour workers. In the conclusion of this book, the socially uneven and often spatially displaced impacts of a sustainability crisis are analysed as 'sociotopic'. Water sustainability crises are good for thinking about the connections between sociogenic and sociotopic dimensions of sustainability crises.

A year after the epidemic, sand-filtration had been built, so effectively eliminating the risks of an equivalently explosive epidemic. So, alongside and complementary to the sociogenic account of 'death in Hamburg', there is, equally necessarily, a biological and ecological account of the dynamics of a bacillus, its natural aquatic environment and human biology. Neither account works without the other. And water as an object of study constitutes the confluence where flows between economy, society and nature meet, on the one hand, and an instance of the complementarity of social and natural science understandings, on the other. While the nineteenth-century miasmatic and contagionist understandings of diseases causally contributed to the sociogenesis of the epidemics, it was the cholera bacillus that delivered the fatal biological blow. People did not die *directly* from their understandings or misunderstandings of the bacillus, but from physically ingesting it, usually with an absence of any knowledge of the biology of cholera, let alone the medical controversies that swirled around them. The epidemics were the outcome of an interaction between the sociogenic and the naturogenic, neither causal series being reducible to the other.

Economy–society–nature

The preceding section highlights a risk of being submerged in complexity. While certainly not proposing one right way, as a starting point this book draws on the work of Karl Polanyi's economic anthropology in the way it asks questions of, and analytically frames, water both as an object of study and a probe into society–nature interactions. In so doing, it develops the approach adopted in *Exploring the Tomato* (Harvey et al., 2002), where the historical transformations of the tomato were analysed to understand the emergence of mass production and consumption in capitalist economies, culminating in the supermarket systems of the present day. In contrast to the tomato with its fully commodified trajectories in production, markets and consumption, water as a good is much more varied and unstable in the economic forms it has assumed. And, also in contrast to the tomato with its hybridisation, genetic modification, and sometimes quite revolutionary innovations in cultivation, domesticated water presents a quite different, more crisis-ridden, history of society–nature interactions – droughts, epidemics, and a range of risks to societal sustainability.

When addressing water as a focus of interconnection between many different societal processes, as well as society–nature interactions, there are many different angles of approach. The figure below illustrates the one adopted here, putting economies, in this case economies of water, at the centre of its analysis, and working from there to explore interactions between economic, cultural, political, legal, biological and environmental processes. It cannot be overemphasised, however, that adopting this point of entry, and placing the economic figuratively at the centre, is in many ways quite arbitrary, and is not intended to suggest, let alone presuppose, a privileged place for the economy in a nexus of complex interactions, in any spatial or historical context. At most it is suggesting that economic processes are a necessary part of the interactional mix, and, as Polanyi suggested, contribute to sustaining the livelihood of humanity (biological and societal) – a case that is relatively easy to make for economies of water.

The notion of 'economy' developed here is an anthropological one, that is to say, on the one hand investigators carry a concept of economy (transformed and developed over centuries) in their intellectual knapsacks, and on the other, they attempt to project as little of the presuppositions of their own societal economies

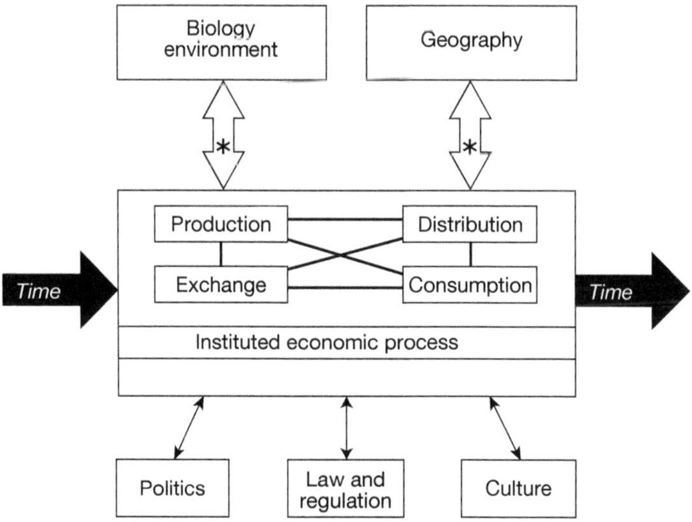

Figure 1.1 Instituted economic process in a complex nexus of interactions.

onto the economies, contemporary or otherwise, they are investigating (Peck, 2013). Thus, in *Aristotle discovers the economy* (Polanyi, 1957), Polanyi argued that a concept of what constitutes 'economy' emerged in Athens at an historical point when commercial trading and price differentials appeared in an as yet embryonic form of economic organisation, since when, of course, both concepts of economy and economic organisational forms have gone through many different trajectories of development. He also argues that, in analysing societies, including those that have no conceptualisation of 'economy' and where economic institutions are not highly differentiated from non-economic societal institutions, understanding of those societies is advanced by having a concept of the economic. We need to be able to answer the question: 'What do we mean by economy?' in such a way as to be open to all historically occurring manifestations (Polanyi, 1966, 307; Harvey, 2007a). So, for example, living in economies dominated by markets and commodities, an anthropological concept of economy needs to distance itself from any presupposition that identifies economy with the market and production for the market, let alone project such presuppositions onto all economic activity, contemporary or historical. Yet, at the same time, current historical conceptualisations, open to further transformation, provide a powerful epistemological platform, or standpoint, for comparative anthropological analysis. So, again for example, from a viewing platform that has seen conceptualisations of 'the economy' co-evolve with the emergence of industrial market capitalisms, the anthropologist is afforded a perspective that enables a more powerful analysis of the difference between varieties of market and non-market economic organisational forms. And for Polanyi, this also means being able to compare and contrast what he called the 'shifting place', or explanatory significance, of the economy in society. Given the earlier discussion of water as both an uncooperative commodity and an unruly public good, this anthropological stance with regard to 'the economy' enables us to approach 'economies of water' enriched by the human historical legacy of both conceptual and economic variation.

To advance this anthropological stance towards 'the economy', Polanyi himself proposed a minimalist analytical framework for exploring societal variation of economies, in the sense of one minimally constrained by presuppositions of any actually existing economic organisation. He proposed two – and only two – processes, which constitute an economy: processes of displacing entities in space, and processes of changing who exercises control over those entities (Polanyi, 1957; Harvey, 2007b). Neither of these two processes are intrinsically economic, and only become so when organisationally linked to each other – or, in other words, socially instituted in relation to each other. He broadly distinguished between three main modes of instituting these processes in relation to each other, in predominantly kinship societies through reciprocation, in archaic and monarchical societies through redistribution by central authority, and in market societies through monetary exchanges. The anthropological openness of this concept of economy is what enables it to encompass a wide range of societal organisation of how goods, people or activities are moved about in space and change the hands that control them. Economies do exist and have existed without money, individual greed, rational calculation or tradition-breaking innovative growth. Moreover, the openness of the concept makes no presupposition that these different modes of instituting the economic organisation of these processes are exclusive of each other: societies can combine, in many different ways, reciprocation, redistribution and market exchange – a modern welfare state together with contemporary household economies would count as an example.

Yet, Polanyi's concept of economy is perhaps a bit too minimalist, and in restricting his concept to that of a two-process economy reduces the analytical capacity to make some important distinctions between historically occurring societies. The concept of economy developed here, and in previous work, adds a further two processes to Polanyi's two: processes of transformation of qualitative characteristics of entities (objects, activities, people); and processes of transformation of the functionality of using those entities. Again, these processes can only be considered economic relationally rather than intrinsically when they are in socially instituted configurations in relation to the other three processes (see Figure 1.1). In broad terms, and again with very little presuppositional baggage, these additional processes can be seen, in economies familiar to us, to correspond to processes of production and end consumption (by the final consumer). But, for example, processes of transformation of qualitative characteristics can embrace anything from the most elemental preparation of food to producing high-tech electronic equipment. By adding these dimensions – and without going into further depth or detail here – we are able usefully to understand processes of economic transformation which entail more than a change in how goods are moved about in space or transfer ownership or control. For example, the Spanish encountered lively markets for tomatoes in sixteenth-century Mexico. But a totally different economy of tomatoes developed with the industrial production for mass consumption of Heinz tomato ketchup or Campbell's tomato soups in late nineteenth-century America. With these two additional processes we can explore variation in *what* is moved about or changes hands, and *how* people consume.

So, now, when exploring economies of water we have an anthropological concept of economy that grasps how water is produced, distributed, controlled and consumed in different organisational configurations. And, following Polanyi, we make no presumption about societies being characterised or dominated by a single organisational configuration, indeed, we can anticipate combinations of different forms of economic configuration. Jumping forward, for example, there may be public provision of tapwater combined with market commodity provision of bottled water, under two very different processes of production, distribution, transfer of ownership and patterns of consumption.

Adopting this anthropological stance towards economies of water in respect to four transformational processes provides a more integrated understanding of economic organisation. If we take processes of changing the transformation of the qualitative characteristics of water, there can be whole sequences of transformations, from digging wells or extracting river water, to processes of settling, filtration and chlorination, adding or subtracting minerals, boiling in the home, adding flavours (tea, sugar, etc), freezing, and so on, prior to final consumption. Glucksmann's concept of consumption work highlights the phase of the production process that occurs *after* the final transfer of a good, and, whether in a market or non-market society, or a market or non-market good, that final phase is critical in deciding what, qualitatively, is consumed (Glucksmann, 2009, 2013). That is certainly the case for producing water as it is to be consumed: plain tap, filtered, cooled, with juice, in tea, coffee, soup, etc. An anthropological stance takes the sequence as a whole as the production process, rather than identifying production as that which is undertaken by commercial actors or public authorities. The question is how production is organised in a configurational relation to exchange, distribution and consumption. To take the activity of producing a tea drink in Taiwan, for example, bottled tea contrasts with tea brewed in the home and then chilled in the refrigerator. The qualitative transformation of water occurs in a different relation to how water is distributed via public tapwater infrastructures or commercial bottle retailers, whether a particular qualitative transformation occurs before or after the water changes hands, and last but not least, how and where tea is consumed. Moreover, as with real or instant coffee, it can be argued that bottled and domestic chilled tea are qualitatively different drinks, resulting from the different transformation processes and material inputs. As Glucksmann has argued, there are historically shifting divisions of labour between consumers and other provisioners in the realisation of a good for final consumption.

There is a similarly wide range of distribution processes for water: private domestic borewells, extraction from springs, piped infrastructures in big cities, bottles, tankers, street vendors, to name but some. As with production processes, these different modes of distributing water are articulated differently with how water is made fit for human consumption, under whose control different phases of the distribution processes occur, and how water is finally consumed. Again, from this anthropological stance the distribution process is seen as an integral process, from 'source to sip', where the final phase may be drinking direct from a street water fountain, a glass filled from a tap, or domestic or office bottled water dispensing machine, or, indeed, a water-bottle nipple.

The process of transferring control, already discussed at some length above, needs to be seen within this 'instituted economic process' (IEP) perspective to include the whole trajectory of source to sip, often entailing 'changing hands' of very different kinds along the way. Even at the end point, in the domestic household, there may well be differential rights to drinking water resources and who gets how much of what (soft drinks, bottled water, etc.). One of the key issues is how and whether ownership of naturally occurring water resources (groundwater, rivers) has been instituted. By analysing the effective control and its transfer of the water resource as the economic process, the approach enables analysis of formal and informal, legal and illegal transfers of control. So, for example, millions of middle-class households in Delhi appropriate publicly owned groundwater illegally by digging unlicensed borewells on their domestic properties. The integrated approach moreover highlights aspects which may be overlooked: public water infrastructures and provision of tapwater by a public authority is generally complemented by privately owned domestic infrastructures, and the interface between the two (the transfer of control) historically has often been highly contentious. In nineteenth-century London, for example, private water providers were given powers by public authorities to inspect and impose modifications on private domestic properties to limit risks of unmonitored water wastage (the UK is quite atypical for its extremely low coverage of water meters) (Taylor and Trentmann, 2011; Trentmann and Taylor, 2005). In many places in the world, there is a combination of public water supply and domestic private ownership: the household has private ownership of the water once it crosses a physical boundary. The IEP approach therefore embraces the successive transfers of control between different social actors, in order to grasp the overall economy of water.

Finally, how water is consumed is critical for the instituting of any economy of water, including when adopting a specific focus on ways of drinking water. Viewed within the IEP approach, consumption is both a process in its own right, and, relationally, the process which both ends and initiates the cycle of production, distribution and exchange. In that sense, the process of consumption and its repetition is what ensures coherence and reproduction of the relational configuration. People consume water, including the diverse ways of drinking water, and then go back for more; and frequently, although not always, more of the same. The force of norms and routines of consumption, discussed below, is thus critical to the formation of economies of water – and for any instituted economy, for that matter. In this analytical framework, therefore, there can be many ways of using water that are not considered as 'consumption': using the sea to swim in; cooling down by taking water from a stream; making and throwing snowballs; the occasional drinking of rainwater. As an analytical choice, the using or using up of water becomes consumption, an economic process, only when in a relational configuration with other processes, normally entailing some level of social differentiation in the undertaking of processes. In other words, the empirical starting point is not the mythical abstraction of Robinson Crusoe, single-person, self-subsistence survival, inhabiting a state of nature environment. Rather, if there are any starting points, they are the myriad of historical social organisations of earlier configurational forms or producing, distributing, appropriating and consuming water.

Adoption of this relational and configurational approach sidesteps the irresoluble paradox which emerges in some economic literature relating to market–commodity societies, and which relies on assumptions of a radical causal or agency autonomy of production and consumption. Are consumers only the choosers from the array of goods designed by producers, or are consumers, partly through their choices, driving the emergence of new products, or the persistence and popularity of existing ones? The question of whether producer or consumer is 'first mover' is recast when both are always 'moving' from within existing relational contexts. As already referred to, over the recent decade Taiwan people drink more bottled tea than bottled water, but both producers and consumers were already inhabiting an economy and culture of tea drinking. Germans overwhelmingly drink highly mineralised sparkling water at table, in preference to tapwater or still bottled water. But this emerged from a prior context of an extensive culture of spas and 'taking the waters', often also highly mineralised (Back et al., 1995). This can neither be explained by consumers independently of each other one day all deciding that sparkling water is best, nor of producers, also independently of each other, suddenly coming up with the idea to create a market for sparkling water. Rather, this particular economy of bottled water emerged out of historically preceding relational configurations of consumption and production, characteristic of Germany, just as in the case of tea in Taiwan. On a larger scale, the emergence of tapwater infrastructures with sand filtering and chlorination has emerged, at least in Europe, with different trajectories, and different economic organisation – as already illustrated by the contrast between Hamburg and Altona. There is no temporal ground zero, where either producers or consumers independently of each other and of their previous configurational context, initiate a novelty.

This relational and configurational concept of consumption thus places consumption processes on an analytical parity with the other three processes, and in so doing accords them a central importance often missing in economic analysis. So, firstly, as an anthropological take on consumption, it permits comparison with non-market, non-capitalist, and non-commodity forms of consumption. Secondly, within market economies, it also enables analysis of consumption of public goods, without reducing all provision of public services and their use by citizens to market provision and commodity consumption, even when current political rhetorical devices and discourses may do so. And thirdly, it emphasises how much *either* market *or* public provision of a good relies on the non-market activities of the consumer, not only in their completing the process of production to the point which defines the quality of the water that is actually consumed, but also the manner in which it is consumed, notably the social context and temporalities of consumption.

Thus, by way of illustration, let's take two stylised examples of drinking tapwater and drinking bottled water in the United Kingdom. Consumer practices and routines of drinking are critical in defining how, how much, and what end-quality of water is consumed: 'raw' and 'neat' or as tea, coffee, squash or soup, with meals or at routinised breaks. Domestic drinking times have some quite rigid habitual patterns, cups of tea/coffee first thing in the morning, then often mid-morning, after lunch and so on. People have their water drinking schedules, domestic, at leisure and at work. The institution of tea-breaks in workplaces illustrate how collective and social such consumption practices are, and indeed, they may be the outcome of union struggles for a right to welfare (Holgate, 2005; Hayes, 2002). The consumption of bottled water, and the design of different size and shaped bottles, presume rather than script the multiple ways in which consumers consume water on the move or at home, fitting in with sporting activities or commuting. How the consumption of bottled water articulates with other social practices is an integral part of the coordinating activity of consumers, even in the case of highly individualised or idiosyncratic routines. As we will see, processes of consuming water vary enormously, even between societies with similar public infrastructures or with markets for bottled drinks. But that societal variety is itself one of routines and norms, and, even when novel routines and norms emerge as they often do, they do so not from a prior state of normlessness, but from previously configured combinations of routines for drinking water. In the IEP perspective, varieties of consumption of water for drinking are co-constitutive of the configurational variety of economies of drinking water in Europe, Delhi, Taiwan and Mexico City, which will be a focus of our analysis.

Now, having attempted an answer to Polanyi's question 'What do we mean by economy?', let us turn to considering the place of the economy in the complex array of interactions represented in Figure 1.1. As stated above, 'the economy' and economies of water are only a chosen point of entry into a complex of interactions, two of which will be of particular concern: economy–geography interactions and economy–nature interactions, the arrows starred in Figure 1.1. These two highlighted interactions are obviously intimately related. Economy–nature interactions occur in a geography of natural environmental settings. Analytically, economy–geography interactions concern the economy's use and appropriation of the natural endowments (water resources), whereas economy–nature interactions focus on biological and natural sustainability, and the impacts of human activity on natural processes, both biological and environmental.

There is a general case that historical trajectories of societal economies are deeply conditioned by their natural resource environments, and the finitudes of those resources (Harvey, 2014a). Exploring and comparing historically developing economies of water and their natural water resource environments provides a particularly fruitful way in to understanding the evolving interactions involved, their scale and the different resource endowments. Thus, urbanisation in nineteenth century Europe was marked by a radical shift from the use of groundwater as a resource for domestic water to the use of rivers as the dominant natural resource, a shift both in scale and in the nature of the resource. Shifts in use of water as a resource for production of domestic water, however, may destabilise historic forms of appropriation, namely, who, what social entity, has rights to particular water resources. As we have already noted, the historical development of scales of appropriation can be quite higgledy-piggledy, gradually emerging into more or less clear local, regional or national rights. In the United Kingdom, the English right to Welsh water was hotly contested, and the development of appropriation rights over water resources at various scales is often a turbulent one, Manchester with its nearby rural and mountain districts and jurisdictions, for example (Sheail, 1986; Hassan, 1985, 1998). As we will see in the case of Delhi, the upstream riparian states have been in conflict with the capital's claim on the river Yamuna when constructing upstream dams to increase water resources, and the downstream riparian states are in conflict for the economy–nature interactions resulting in the city's extreme pollution of the river.

Turning to economy–nature interactions and sustainability, the cameo example of nineteenth-century Hamburg demonstrated the impact of human activity and migration on the local river water resource resulting in a quite particular sustainability crisis: a cholera outbreak with a distinctive Hamburg signature. Again, a general case can be made for analysing sustainability crises, including climate change, as emergent from particular economy–nature interactions, as sociogenic rather than anthropogenic (caused by an amorphous mankind in general or in aggregate) (Harvey, 2014a). The IEP approach is one that situates economies in their natural environments and their extended rights over global resources, rather than capitalism (tagged neoliberalist) in general with planet earth or Nature with a capital N (McCarthy and Prudham, 2004; Heynon et al., 2007). Emergent water sustainability crises, such as that in nineteenth-century Hamburg, will help develop analytical tools to understand the varied dynamics involved. We shall be comparing London with Mexico City and Taiwan with Delhi, for the contrasting dynamics and variety of sustainability crises they have experienced and are experiencing. To take one peculiar example of an emergent and sociogenic sustainability crisis, over-abstraction of groundwater by farmers in competition with industry and domestic users in Western Taiwan has led to land subsidence undermining the high speed train from Taichung to Kaohsiung. That instance is exemplary of a particular sociotopic impact (farmers versus train passengers), characteristic of water sustainability crises: the socially and spatially dynamic relation between generation and impact.

Discussion of economy–geography or economy–nature interactions necessarily raises the issue of the scaling of economies, the environments from which they draw their (water) resources, and the environments on which economies impact. Water always involves distinctive processes of socio-economic scaling. The emergence in the nineteenth and twentieth centuries in advanced economies of the 'municipal hydraulic paradigm', large-scale and integrated urban infrastructures together with new scales of governance (Bakker, 2010; Swyngedouw, 2004), exemplifies a widespread form of scale formation. Scale formation is the outcome of economy–geography and economy–nature interactions, rather than being preformed or naturally given (Thrift, 1996). And, historically, scales are transformations of previously instituted scales. Some have argued for a splintering of previously integrated scales (Graham and Marvin, 2001) under the pressure of neoliberalism and privatisation. We draw upon and compare contemporary Delhi with other haphazard and multi-scaled water economies, such as those of Jakarta (Kooy and Bakker, 2008). There have also clearly been contrary movements, with river basins emerging as a distinctive scale of economic, environmental and governance organisation (Hassan, 1998; Bakker, 2003b). Yet, scales also have a time duration, and it could be argued that some scales remain frozen in time over quite long periods, resistant to re-scaling. By contrast, the South–North Water Transfer Project in China, which embraces a developing complex of levels of governance incorporating 22 cities as well as Beijing along the system of canals, presents a stunning case of scale formation occurring within a distinctive and changing political economy (Wang and Huang, 2013). The configurations of economies of water, in an IEP perspective, thus provides a rich resource for analysing processes of scaling, rescaling, and scale durabilities as immanent in the complex of interactions represented in Figure 1.1.

This book arises out of research into economies of water in the United Kingdom (with a special focus on London), Delhi, Mexico City and Taiwan, and is both historical and comparative. At one level, we are just following the primary anthropological directive of making the ordinary appear extraordinary by viewing each field of research from the perspective of a different time and place, from outside our conceptual comfort zones. We have to grasp how it is, and under what conditions, contemporary Western tapwater became so normal and so taken for granted. That becomes more difficult to explain when one sees just how many obstacles and challenges exist in Delhi today to achieving that level of normality, even for the middle classes and rich. And we then also need to recover the traumatic histories, whether of epidemics, droughts or floods already alluded to, behind the veil of current normality.

At another level, historical and comparative research has the overriding purpose of stretching and expanding our conceptual frameworks – even to breaking point. It forces us to question what we mean by the quality, the ownership, or the sustainability of water, and then more broadly how we understand the interactions between economy, polity, culture and nature. For example, from our starting point with bottled water, it became very clear that processes of qualification – from source, production, marketing and branding, through to actual consumption – differed markedly within Europe when comparing the United Kingdom, Germany and Italy. The contrasts became even more striking when introducing Mexico City, Delhi and Taiwan into the mix. Then, from this starting point, it became clear bottled water as a quality of drinking water could be better understood both by placing it in the context of other bottled drinks, but more particularly, in the context of the varieties of tapwater in our different cases. Much of the discussion around processes of qualification has been focused on market commodities (of which commercial bottled water is a case in point) (Callon et al., 2002; Harvey et al., 2004; Allaire, 2004; Ponte and Gibbon, 2005). So the long history of the qualification of tapwater as a public good, the continuing regulation regimes of tapwater quality, *its* distinctive quality crises and – to step outside its ordinariness – the peculiarities of all-purpose water in a modern Western city, force us to expand our conceptualisation of processes of qualification.

Likewise, ownership or proprietorial rights over water resources test conceptualisation of property in quite distinctive ways. How different is water from knowledge as historically constituted public goods? Both involve interactions with natural processes, yet, outside the laboratory context or scientific sites of control, scientists make few proprietorial claims over natural resources, quite unlike rights over water resources. By comparing the two fields, understanding the difference between water and knowledge as public goods stretches our concepts of property, both public and private. Comparison and historical analysis of economies of water presents a complexity of forms of appropriation, often found in combination, legal and illegal, formal and informal, and at various scales, the microcosm of a slum in Delhi to national territorial rights over all water resources in Taiwan. One can ask the question as to what kind of property is the rainwater collected in a water butt, in the context of domestic private property (as in England), as against water drawn by a well from an aquifer, also on private domestic property (as in Delhi). In short, water, in all its social and historical variety, is good for thinking about property.

So, to be clear, this book is not a comparative set of case studies, with a purpose of classifying and analysing each of the cases in their own terms. Rather, the variety, both historical and spatial, is used to develop and stretch our conceptual frameworks, what we mean by an economy of water and how we understand polity–economy–nature interactions.

Themes and threads

Each of the chapters of the book highlights particular conceptual challenges, and, with the cumulative succession of chapters, develops the range and reach of the IEP approach.

Chapter 2 introduces the theme of configurational change by tracing the emergence of large scale urban infrastructures in London from the seventeenth century through to privatisation in the late twentieth century. This historical trajectory of successive economies of water, from the late eighteenth century intimately conjoined with economies of sewage, is theoretically used to compare and contrast with transition theory accounts (Geels, 2002, 2006; Geels and Schot, 2007), but also to analyse the contrasting dynamics of reconfiguration at different points in the history, for example, between the sustainability crisis of the mid-nineteenth century (known as the Great Stink) (Halliday, 2001), and politico-economic turbulence preceding the municipalisation of London's water and sewage systems in the early twentieth century. In so doing, this chapter presents and develops diagrammatic representations of economies of water, an analytical tool which is then used for all the other economies of water we examine. The distinctive historical London trajectory thus serves as a reference point for later comparison especially with Delhi and Taiwan, contrasting the endogenous processes of change with two examples where colonisation by the foreign powers of England and Japan respectively super-imposed external technologies and economies of water, with lasting legacies in both cases. The London trajectory also provides an account of the long process of qualification and re-qualification of tapwater, as safe for all purposes, through integrated water-sewage infrastructural design, reservoirs and filter beds, culminating in the universal application of chlorinated water fully achieved only after the Second World War.

The establishment of almost universal potable but all-purpose tapwater in the United Kingdom by the late twentieth century provides the essential background context for the later emergence of dedicated drinking bottled water as fully commercial market goods in the UK. By contrast, in the USA in cities such as New York, a thriving bottled water industry in the early twentieth century was effectively destroyed by the development of municipal, public, tapwater (Back et al., 1995; Davis and Davis, 1997). In France, the much later and more partial coverage of all-purpose tapwater had witnessed a significant bottled water presence (Goubert, 1989), and, indeed, to this day in Italy the lack of trust in purity and safety of tapwater remains a contributing condition for the much larger bottled water market and consumption. So, based on interviews with key informants, survey data, drinks diaries of consumers, supermarket internet and store fieldwork, Chapter 3 analyses and compares the contrasting economies of bottled water in contemporary England, Germany and Italy, with particular emphasis on England. With both Italy and

Germany drinking almost four times more bottled water per capita than the UK, we seek to explain the variation in terms of the configurational differences, the importance of supermarkets in the UK, of local and regional producers for local and regional consumers in Italy, and a distinctively divided Germany between discount retailers and dedicated drinks producers.

The role of European legislation in defining the difference between spring and mineral water plays a role in shaping the qualification process, affecting the quality varieties occurring in Germany in particular. As naturally carbonated mineral water from a spring source can involve many stages of decarbonation, demineralisation and then reconstitution, before reaching the consumer and satisfying the particular cultural forms of quality appreciation involved in actually drinking the bottled water, we see the process of qualification as a continuous, but phased, process 'from source to sip'. In contrast to Callon's own accounts of economies of quality, where the discussion of the relation between price and quality is remarkably absent, we explore the complex relationship in the different market organisations. This three country European comparison then becomes a reference point for exploring the economies of bottled water in contexts where tapwater is not virtually omnipresent, including Mexico City which has the highest per capita bottled water consumption in the world, complemented by almost as remarkable levels of soft drinks.

The next three chapters extend the comparisons to countries and cities where there is no universal availability of immediately potable tapwater either in public or privatised infrastructural forms. Delhi is a city of many economies of water (Chapter 4), some quite localised, with approximately only 1 in 10 households linked to tapwater infrastructures, providing an intermittent supply of dubious and irregular quality. Fundamental rights to water of any quality are unmet in many of the unauthorised slums, partly as a consequence of the unplanned growth with a megacity population now reaching nearly 20 million (Baud et al., 2008; Davis, 2007). We describe the 'bricolage' economies of water (Maria, 2008a, b; Cleaver, 2012), where citizens of the city, whether middle class or the urban destitute, develop coping strategies, putting together water from different sources, formally and informally, legally and illegally, to meet their water needs. 'Ownership' of water in this context, the effective social means of control over water resources, embraces a variety of hybrid and combinations of property rights. Referred to briefly above, over a million private domestic borewells in middle-class homes involve domestic (non-market) private property, illegally appropriating putatively common pool resource groundwater, yet with public authority connivance and complicity, in the absence of any prospect of enforcement of public rights. One slum illegally appropriated water from the public tapwater infrastructure to supply its own population, but then, in an informal market, sold some of its supply to a second slum, which was too distant from the infrastructure to illegally tap into it directly. These are concept-stretching examples of forms of social appropriation and exchange of water resources.

We interviewed slum dwellers in authorised and unauthorised slums, water experts at the World Bank, trades union officials, middle-class consumers, water street vendors and tanker drivers, municipal authorities, bottled water producers and retailers, NGOs and senior officials at the Delhi Jal Board (DJB, the main public water organisation). We supplemented our own primary research with secondary literature not only about Delhi, but about other major cities in India. It is impossible to give an exhaustive account of the multitude and variety of Delhi water economies, let alone urban and rural India, and that was not our aim. The extreme inequalities in rights to water and to sewage infrastructures stretch our conceptualisation of 'appropriation' and exchange as instituted processes. Yet Delhi and the river Yamuna at the same time has generated its own and ongoing sustainability crisis, with the river being described as a sewer in ways reminiscent of mid-nineteenth century London, by a top DJB official. And bottled water for the better off, street vendor water for the less well off, present a distinctive quality spectrum for potable water in the context of polluted and unreliable tapwater.

A shift of place and shift of megacity: the Mexico City metropolitan urban area has been described as a city of many cities, with a population of over 21 million (Chapter 5). The shift of place entails immediately having to grasp the idea of a huge city, 2,200 metres above sea level, and without a large river flowing through or near it as a source for its water. This creates conditions for a quite different economy–nature interaction for its economies of water. But there is also history: the Spanish arrived to find an Aztec capital city (Tenochtitlan) of lakes (fed by the substantial summer rainfall) and islands. They altered the whole subsequent history of water provisioning by draining the lakes. Mexico City (and its surrounding urban additions) now depends on getting its water from deep wells, on the one hand, and ever more distant water resources, including the recently developed Cutzamala system bringing water from 100 kilometres away. This in turn forms the background context for major social conflicts over rights to water resources, particularly between the city and the rural indigenous populations in areas from where water is drawn, and across whose land the water traverses. The absence of proximate rivers also laid the conditions for severe pollution of sub-city aquifers, as well as their over-exploitation for drawing water. So, Mexico City exemplifies distinctive processes of sociogenic sustainability crises: severe land subsidence in the city, water shortages, contaminated deep well water, major cholera outbreaks in the 1990s, and untreated effluent transported into the river systems below, dumped in the food-growing agricultural peripheries.

Restricted to two main water resources, the greater Mexico City urban area, in contrast to Delhi, has much higher levels of tapwater and sewage pipe connection. Although there are irregular and illegal slums, they are not on the same scale. So, although Mexico City has highly socio-economically divided economies of water for rich and poor, it also has three large-scale components: unpotable tapwater, unpotable tanker water and potable processed water in 20 litre bottles. The variation within the urban area and social economic divisions is related to the extent to which people depend on one or other of these water resources, all of which involve different mixes of public, public–private, and international commercial, appropriations and exchanges of water. The water tanker system, for example, involves both private commercial and public providers, and is on a much greater, city-wide, scale, locked in to a domestic plumbing system of large underground cisterns and roof-top water tanks. Likewise, the city-wide bottled water distribution system connects to domestic infrastructures for cooling and storing drinking water, including dispensing machines. So Mexico City extends our comparative horizons by providing distinctive combinations of, and interconnections between, differentiated water *qualities*, socially divided *property* rights to water resources, and sociogenic *sustainability* crises, together with highly distinctive consumption patterns, especially of bottled water.

Colonised and under Japanese military rule from 1895 to 1945, and then under a dictatorship until 1996, the historical evolution of economies of water in Taiwan reflect the centralising force of a developmental and authoritarian state (Chapter 6). From 1901, the Japanese 'nationalised' all Taiwan's plentiful, sub-tropical water resources flowing from the mountains, expropriating indigenous peoples' collective and communal rights. Initiated by the Japanese who built the first urban infrastructures linked to mountain dams, the Taiwanese comparison provides an exemplary case of a polity–economy–nature interaction with the gradual scaling up of tapwater provision, before and following the Second World War, and then significantly accelerated by a sustainability crisis impacting on the Taiwanese economy: cholera outbreaks in the early 1970s. This comparative distinctiveness enables us to explore the dynamics of evolution of a public tapwater system and complementary market economies of water provision. Thus quality inadequacies and interruptions of tapwater supply, stretching into the final decades of the twentieth century, are compensated for by domestic water storage systems, and by the universal practice of boiling water, often also filtering beforehand, even in the capital city. But in the particular conditions of the South West, in Kaohsiung City and Pintung County, where effective tapwater provision of potable quality has been slowest to develop, peculiar markets for filtered water have been instituted, largely drawn from the public water supply. Into the 1990s, levels of bacterial contamination of tapwater, connected to intensive and large-scale pig farming, infected social memory as much as the water itself. There are now over 1000 water 'garages' in the city, parading their high-tech filtration equipment, where people fill up for their drinking water and water for making tea, with a market choice of up to five different qualities of drinking water at different prices. Drinking water garages manifest quite peculiar and, to the non-Taiwanese, extraordinary process of qualification of water. These societal variations in economies of water enable us to deepen our analysis of the interactions between public and market economies of water, as well as distinctive tea and soup cultures of water consumption.

This richness and variety of the different polity–economy–nature interactions involved in economies of water in our comparison between three European countries, Delhi, Mexico City and Taiwan forms the main resource for the concluding chapter of the book. In this chapter, from a comparative and historically evolutionary perspective of polity–economy–nature interactions, we focus particularly on an analysis of the *connections between* the central themes of the book: economies of water and their transformation, quality and sustainability.

The book offers a rich resource for comparing processes of transition and transformation, starting with the IEP configurational perspective. Engaging in particular with transition theory and political ecology, our cases contrast exclusively endogenous change within London's economies of water, with the additionally exogenous change resulting from invasions and colonisation in the case of Delhi, Mexico City and Taiwan. With endogenous transitions, by comparing successive and sequential transitions, it is argued that explanation is necessarily contextual and historical, triggered by internal contradictions and by sustainability crises. It demonstrates scale friction and scale formation between economic, fiscal and political processes, thus problematising the static conception of a pre-given Multi-Level Perspective. The analysis responds to Bakker's ideas of water as an uncooperative commodity, by arguing for an historical account of the emergence of new forms of public, hybrid, and indeed commodity goods in different historical and societal contexts.

Affirming the multi-modal interdependencies between public, hybrid and commodity drinking water in all our cases of economies of water, the IEP analysis develops an integrated and holistic account of the 'economies of quality', processes of qualification of water 'from source to sip'. In this analysis, the significance of the role of consumers in the final qualification of water as consumed attracts particular analytical attention. By comparing tap and bottled water in different societal contexts, conditioned by their multi-modality, it becomes only too obvious that there is no single thing as 'the quality of drinking water', and that the quality of water is always inscribed in a quality–price/cost matrix.

If these comparisons involve historical and societal variation, the final concluding section compares the different society–nature interactions involved both in the growth of economies of water, and in the sustainability crises they generated. The analysis of drinking water exemplifies the sociogenic character of the generation of sustainability crises, but equally the sociotopic character of the impacts of crisis in terms of who, where and how different sectors of society suffer the consequences. Water sustainability crises provide some cases where those who generate the crisis themselves suffer the consequences, and others where impacts are quite remote from their source. For that reason, water sustainability crises provide a rich source for reflecting on sustainability crises more broadly, including climate change.

So, finally, in the current circumstances of climate change and the exploitation of the planet's finite resources, whether of water, land, oil, or materials, social scientific analysis confronts the 'shifting place of society in nature'. The analysis of drinking water demonstrates the need and the challenge to grasp the three-way variation in combination between historical, societal, and society–nature interaction processes. Water is ever a fulcrum of historical change, societal variation and society–nature interactions in particular environmental contexts. Politics of water, at various scales and levels, have been deeply implicated within these dynamics of variation. The book speaks to the need to develop politics-in-context, societal and environmental, at multiple scales, local, societal and international in addressing the challenges of climate change and the finitudes of earth's resources.

2 London drinking

Historical transformations of water

> Water charges discriminate against working people and the unemployed in favour of the wealthy and are another regressive tax taking vital money out of the pockets of people and out of our economy. Our public water system is already paid for through general taxation which is progressive and we wish it to remain that way.
>
> *Right2Water*[1]

Before analysing the long historical evolution of drinking water in London, it is worth noting that history is always viewed backwards from a particular contemporary historical standpoint. This chapter is a history from a contemporary British historical standpoint. History appears quite differently from an Irish historical standpoint, and so shines an initial contrasting light on the specificities of the British, so emphasising that this chapter is of a particular and contingent historical trajectory. The Right2Water campaign in the Republic of Ireland burst onto the streets with mass protests and demonstrations in 2014, demanding water as a free public good, invoking the United Nation's designation of safe drinking water as a human right. Ireland is unique among OECD countries in having no direct charges related to the provision of drinking water, which historically has been financed out of general taxation. In order to modernise water provisioning and sewerage, the government had set up a company, Irish Water, a subsidiary of the state-owned gas utility company, Bord Gáis, under the Water Services Act, 2013. Irish Water is a national-scale organisation, replacing the previous responsibility resting with local municipalities. The need for radical reorganisation was said to arise from the dilapidated and deteriorating state of the infrastructure: the pipework for water and sewage, and related technologies and infrastructures for water and sewage treatment. At the same time, a radical change in financing the water and sewage provisioning was to be achieved through the introduction of water meters, and charging by volume of usage. This contemporary political and financial upheaval illustrates in a striking manner the potential for conflict and contestation over what constitutes a public good, how it is paid for, what

1 http://www.right2water.ie/ [access date 13 August 2014].

economic organisation is involved in delivering it, and at what scale. Suddenly, all these interconnected elements were destabilised in an advanced economy and a water-rich environment. Is water only a public good when a nation state finances it out of general taxation and provides it 'free at the point of delivery' (as with health services under the UK's National Health Service)? As we shall see, this is a peculiarly Irish conception. But it highlights how the construction of water as either a public or private good evolves in different ways in different societal contexts.

This chapter examines the co-evolution of qualities of drinking water, its nature as an economic good, the technologies of its production, the process of spatially scaling up its economic and political organisation, even the changing patterns of consumption. London is taken as an empirical case of a particular evolutionary trajectory, although many aspects were and are shared by other industrialising cities in the United Kingdom (Hassan, 1985, 1998). Employing the IEP configurational analysis, the aim is not to provide a comprehensive history, but to focus on processes of major transitions, crises and the dynamics of change. The outcome of this history, as a lived reality of how people drink water from taps in their houses, appears almost self-evident and inevitable as the only way of both providing water to mass urban populations, and disposing of their waste waters. Provision of 24/7 piped water under pressure connected to the domestic infrastructure of every dwelling, suitable for all purposes from drinking to flushing toilets, cooking to washing cars, cleaning bodies in baths or showers to watering gardens, has been given the academic imprimatur of the 'municipal hydraulic paradigm' (Bakker 2003a, 2010; Swyngedouw, 2004). Yet, the history we examine here is marked by crises and instabilities, conditioned by a specific political and environmental contexts. There is no possible teleological account of an endpoint that was always meant to be. The history of London's drinking water guards against thinking that the United Kingdom's water provisioning system has arrived at an endpoint – with a private company, Thames Water, first bought in 2001 by German group RWE, a global business of water provisioning, and then sold on for £8 billion in 2006 to an Australian bank, the McQuarrie Group (through Kemble Water Holdings Limited). And the Irish example serves to emphasise that even such a basic feature as to how water is paid for and financed varies in different contexts. If Ireland is unique, so in many ways, compared with continental Europe, is the United Kingdom, still largely un-metered, so paying for an unspecified and hugely variable quantity of water consumed by different households. In showing how 'economies of water' are instituted, de-instituted, and re-instituted over a long duration time scale, with configurations of production–distribution–exchange–consumption emerging only as transformations of their peculiar antecedents, we begin to develop an analysis of processes of historical, environmental and spatial variation which will serve as a basis for subsequent comparison for other European countries, Delhi, Mexico City and Taiwan. If nothing else, this history shows just how long it took for London to arrive at its particular variant of a 'municipal hydraulic paradigm'.

The tortuous emergence of urban tapwater and sewage, 1600–1990

It was only well after the Second World War, even the 1970s, that baths and indoor toilets became universal to households in the United Kingdom, and only in the mid-1950s that the standard of chlorinated safe water reached universal coverage. These developments took at least two centuries (depending on fixing an arbitrary starting point). Employing the IEP configurational analysis, it is possible to identify five major configurations and the transitions between them. In broad terms these are identified as the pre-industrial (1615–1800), the early industrial (1800–1854), the post-crisis water-sewerage configuration (1854–1904); the integrated municipalised configuration (1904–1989); and the era of privatisation (1989 to date).

The pre-industrial water-sewage configurations

In the pre-industrial period, London as a growing mercantile, craft and financial city already faced significant challenges to expand its supply of water to a population of 250,000 (Hassan, 1985, 1998). During this period there was increasing reliance on river water as a primary source, yet, at the same time, water from springs and wells continued to play a significant role. One of the most exceptional developments, which began to distinguish innovation of water supply and distribution in England from Europe (Goubert, 1989), was the system introduced by the New River Company, formed in 1614 (Graham-Leigh, 2000; Ruddan, 1985). A 40-mile distance was covered by 10 rows of wooden pipes made with 9ft lengths of hollowed-out tree trunks, totalling 400 miles of pipes, delivering mains water supply to London. The company served 55,000 houses with this amazing infrastructure by 1800. Technologically, these pipes running from a source in the river Lea in Hertfordshire to central London, relied on an engineering feat of gravity-fed water over long distance, a significant development from the open conduit systems found in many towns and cities. Houses could only be supplied to ground and basement level. As Ruddan has pointed out, these early water charter companies were typical forms of enterprise where there were no clear boundaries or economic concepts of public versus private. The New River Company, formed by royal charter, was initially part-owned by shares held by the king, as well as being obligated to provide a public utility and given legal protection as such. Supply was intermittent, often one day off, another on, and, as there were no means of measuring volume, the exchange with householders was based on an annual rate – instituting an enduring mode of paying for water that was to last and distinguish the United Kingdom right through to the twenty-first century – a remarkable example of an instituted exchange process, even if it subsequently underwent significant modifications.

Another interesting feature of the New River Company was its mode of appropriation of water, in this case, from the river Lea. Given rights to divert water by royal charter, the company was immediately immersed in conflicts with the interests of bargemen transporting corn into London, on the grounds of interference

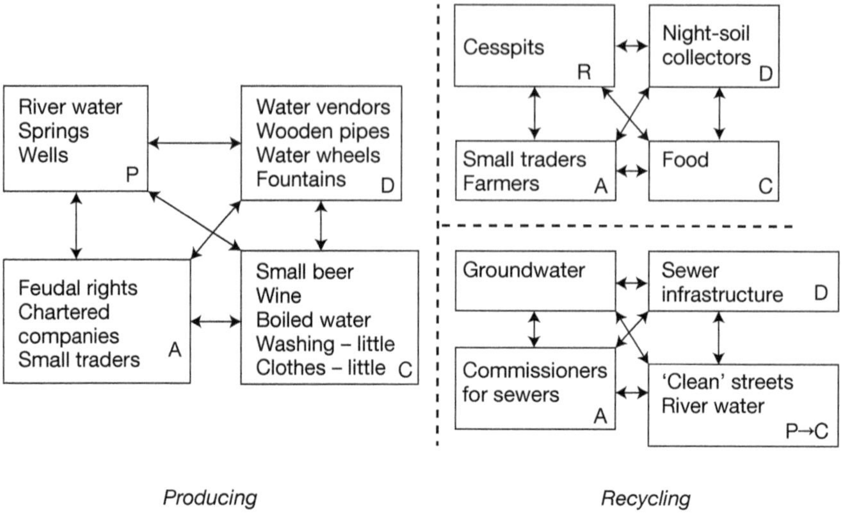

Figure 2.1 Pre-industrial configurations of water and sewage.

by damming and reducing water levels. Legal disputes for extraction and diversion of river water also applied to mill-owners, but by 1670, the New River Company fully secured its rights to extract 'public' water, and then sell it to householders in a form of market exchange. Further legal issues arose concerning the property rights of the land over which the pipes were laid, and this too led to a distinction between a general public right of way, and one limited to the distribution of water. So the company appropriated the water and the rights of its distribution to an end exchange. Initially, moreover, and given the investment in infrastructure, the company obtained effective monopoly rights for distributing water to households at the terminus of the pipes. Finally, householders themselves had to supply the lead pipes connecting to the wooden mains. This early interconnectivity between private domestic infrastructure and external supplier represents an historical starting point of what was to be a focus of tension and conflict: the connecting technologies between the domestic and the external infrastructures, and the fixing of the infrastructure property boundary between the domestic internal and the external. Yet, from the outset of piped systems of provision, there was a critical interdependency between investments inside as well as outside the household, and between private domestic property rights and those of the external provider, whether private or public. From an IEP perspective, in this early case, domestic private infrastructures were linked to a semi-private-commercial infrastructure, which in turn obtained rights of commercial ownership of a public resource (the river Lea), through public license. Although, as we shall see, the company changed character many times, it existed continuously as a legal entity from 1614 to 1904, when it was municipalised and taken into public ownership by the local state.

Only wealthier households, and at that time a small minority of them, had water delivered by pipe-mains or conduit. Wells and springs remained significant, and a water wheel also raised river water from the Thames. Water was widely delivered to households by water vendors, sold by volume – a system which in Paris was significant for drinking water right up to the end of the nineteenth century (Goubert, 1989). In this pre-industrial period, then, there were many sources of water for many different uses.

Evidence of consumption practices at this time is hard to come by, particularly for any but the rich. Generally, clothes and human bodies were washed very infrequently, if ever. Water was certainly used for cooking and boiled for sterilisation. It is unknown how much water as such was consumed as a drink, probably not much at all except by the very poorest. Small beer – a low-alcohol drink requiring boiled water and fermentation – was a more usual mode of drinking water. Progressively, from the eighteenth century onwards, drinking tea with sugar became the 'respectable' way of drinking water, again only for the better-off, along with the increasing fashion of coffee in coffee houses (Mintz, 1985; Austen and Smith, 1992). Drinking water from wells and springs, or using such water for cooking, relied on the purity of the source, and in urban contexts, this became increasingly unsustainable given the wide range of activities polluting the groundwater and shallow aquifers, including cesspits.

The water economic configuration, the left-hand section of Figure 2.1, was thus a combination of different sources of water for different uses, embracing several modes of ownership and distribution, organised exchanges and distinctive pre-industrial consumption uses. This configuration was complemented by two others: sewage and groundwater from rainfall, each with their distinctive instituted economies. Although at that time not economically interdependent with the water configuration in any significant way, they were materially linked in that both sewers dealing with surface water and cesspits with human waste risked serious pollution of river and groundwater respectively. In that respect, they formed a significant precondition for subsequent configurational transitions, which brought these different elements into a single industrial water-sewage configuration. Water fit for many consumption uses required an economic organisation integrating both water provision and waste systems, a requirement often not fully reflected in historical accounts (e.g. Halliday, 2001) or transition theoretical perspectives (e.g. Geels, 2006).

Unlike Moghul India with its dependency and utilisation of monsoon rains, rain and surface water in London were not regarded as a major source of water provision, but rather, in an urban context, as something which required channelling away to ensure streets were navigable. During this period, sewers were gutters, surface channels, and came under the authority of the Commissioners for Sewers, paid for by household rates according to property values. They were regulated by propertied classes represented in the vestries, the local government organisation. Significantly, until 1817 regulation proscribed the dumping of waste or depositing of human excreta into the sewers, although enforcement was meagre, and moral norms less than universally effective. Stretching terminology slightly, the 'consumption' of the public good of navigable streets ensured by this economy of surface water was valued in that it made moving about in narrow streets both easier and more pleasant.

The cesspit economy, by contrast, was much closer to a market economy, but with a scope and scale limited by the dynamics of that economy. In this system, wealthier households with domestic privately-owned cesspits paid nightsoil collectors to empty the cesspits regularly and transport the waste to farms at the periphery of the city, where it was sold to farmers. Farmers, using the manure, in turn produced food for the capital. Consumers purchased and ate the food, completing the virtuous economic circle. To some extent, this cesspit system was to endure until the end of the nineteenth century, although with massively diminishing scope. As the city expanded, the cesspit economy came under increasing strains. Population density and growth of population made cesspits increasingly difficult to maintain without 'negative externalities' to groundwater. Furthermore, as the city expanded, the distance for transporting human excreta grew, placing strains on transport costs. Increasing proportions of the population were failing to pay nightsoil collectors, so adding to the expanding risk of pollution of groundwater. By 1810, there were 200,000 cesspits for a population of one million. In short, the cesspit economy was leaking away before a social, technological and economic revolution sunk it. Many attempts have been made to revive a market economy for human waste in Britain, including famously by Chadwick (Halliday, 2001), but, until very recently, to little effect.

The early industrial water-sewage instituted configuration

The pre-industrial configurations represented in Figure 2.1 were, as suggested, relatively economically disconnected but materially interdependent. The dynamics of transition to a new instituted economic configuration, however, demonstrate just how significant this interdependence was to prove. The driver for change was undoubtedly the rapidly expanding population of London, which doubled from just below one million in 1800, to just below two million in 1850. The entire mode of production of water underwent major changes: the predominant source of water became rivers, the Thames above all, followed by the Lea, the Fleet and lesser rivers. The treatment of river water progressively developed, with the introduction of large-scale infrastructures of settling reservoirs, and multiple sets of slow-sand filter beds extracting organic matter. In short, extraction and transformation of the quality of water achieved an industrial scale.

Secondly, there was rapid expansion of the new technology of cast-iron pipes, replacing conduits and wooden pipes, which permitted the distribution of water under pressure using steam pumps, so releasing water from the grip of gravity. The consequence of this innovation for domestic infrastructure and domestic private plumbing systems was profound, causing tensions at the interface between the general and domestic infrastructures for most of the century. Commercial firms were given rights to enter and inspect domestic installations and enforce adequate provision in successive Metropolitan Water Acts, as the system changed from intermittent gravity-fed supply to continuous secure supply, so removing the need for extensive domestic storage.

Thirdly, piped water under pressure facilitated the rapid expansion of the water closet as the dominant means of removing human waste: water provision and waste removal were technologically integrated, under a new economic configuration.[2] From 1817, the old proscription of dumping human waste into sewers was replaced by a prescription that all new installations of water closets were to be connected to the sewers, the waste now going directly into the rivers. This combination of law and new technology constituted the basis of an emergent sustainability crisis. As Thomas Cubitt graphically expressed it in 1840:

> Fifty years ago nearly all London had every house cleansed into a large cesspool… Now sewers having been very much improved, scarcely any person thinks of making a cesspool, but it is carried off at once into the river: the Thames is now made a great cesspool instead of each person having one of his own.
>
> (cit. Halliday, 2001, 35).

In addition, water closets undermined the economy of human waste: it literally went down the pan; there was no collection of human excrement for manure. Public sewers and public water resources were channels for removal of waste. Human excreta ceased to be a market tradable good, and there was no longer brass where there was muck.

These technological transformations of production and distribution were marked by economic turmoil combined with intense political activity around the emergent new configuration. Prior to the introduction of cast-iron pipe infrastructure, companies were established with local monopolies in settled boundaries by parliamentary statute. Competition between old infrastructures, and stone but ultimately iron pipes, intensified in the second decade of the nineteenth century, with the old wooden pipe companies (the Chelsea Company, York Buildings Company and the London Bridge Company) confronting the new entrants with new technologies (South London, West Middlesex and East London Water Works). Competition was vigorous, with mutual sabotage of pipes, strident negative advertising about price and quality of water from competitors, and several companies digging up the same roads to supply the prime residences. This in turn led to major friction between the public authorities, the vestries and Commission for Sewers, and the private water companies. Competition became so intense as to become self-destructive, eventually leading to deals between companies to swap infrastructures and re-establish local monopolies. The public response to this was the emergence of a 'consumer movement', the Anti-Water Monopoly Association, active through the decades prior to 1850. Usually a powerful advocate of free markets, even John Stuart Mill campaigned for water to be a public utility, on anti-monopoly and common basic public goods grounds (Schwartz, 1966). The negotiated outcome, engineered through the 1821 Parliamentary Select Committee,

2 The water closet with flushing mechanisms was first patented for commercial purposes by Alexander Cummings in 1775, and Joseph Bramah in 1778, who had sold 6000 by 1797 (Halliday, 2001).

was to establish strict private monopolies, but under continuous regulatory monitoring and quality regulation leading up to the 1852 Metropolis Water Act. A deal was also struck, but inadequately enforced, to regulate any potential price rises – an all too familiar story now with a long historical tail.

In terms of consumption, the increasing domestic availability of 'all-purpose' water provided a precondition, but no more, for the emergence of a wide range of new consumption processes, as well as a transformation of the consumption work of storing, treating, filtering and carrying water within the domestic context, as well as from a water source to the household. For example, new elaborate filtering units with charcoal filters graced Victorian households during this period (Hartley, 1964). Again, there is little solid evidence concerning how much water was drunk straight from a domestic tap. Tea drinking became much more widespread during this period with the growth of tea imports from the new plantation economies in India and Sri Lanka (Austen and Smith, 1992). Washing clothes became more of a domestic activity undertaken by servants, partly a consequence of the new availability of industrially produced cotton fabrics. This period also witnessed the emergence of municipal baths supplied by the iron pipe system, for both laundry and washing bodies – a new morality of bodily hygiene marking a significant shift in consumer practices (Crook, 2006). In contrast to France, where water carriers, drinking from separate public fountain sources, and extensive clothes and body washing in the river Seine on elaborate floating boats continued until the twentieth century (Goubert, 1989), the all-purpose piped tapwater in England was largely embraced by burgeoning domestic consumption practices.

Figure 2.2 portrays the newly emergent instituted configuration of production (P), appropriation and exchange (A), distribution (D) and consumption (C). As indicated above, not all the elements of the old configuration disappeared in London, let alone across the United Kingdom. The process of instituting a new configuration progressively enlarged in scale, and was replicated in other urban centres to varying degrees, taking much longer in rural areas. In terms of the processes of instituting this new configuration, some key aspects of the dynamics can be highlighted. Firstly, there was undoubtedly an economic dynamic of new

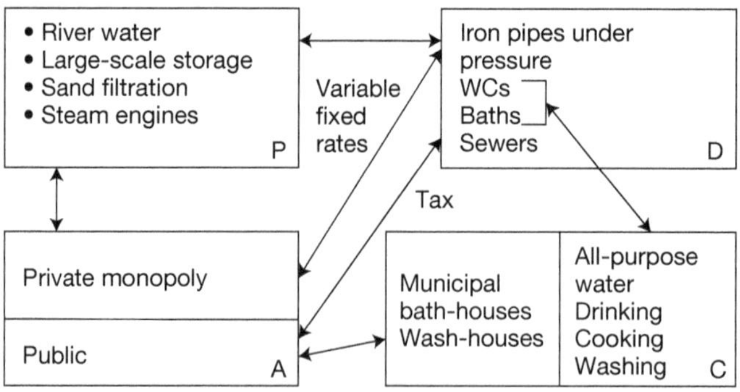

Figure 2.2 The first industrial economic configuration of water and sewage.

entrant firms seeking profits by expanding market share, and deploying new technologies of water extraction, transformation and distribution to do so. In terms of demand, there was an expansion of demand arising from an exploding urban population, but also convenience, safety and quality of such a good reflected an enduring need, which even at that time was considered as 'basic'. Configurational change entails complementary dynamics within each of the four interrelated processes, involving innovation in consumption practices parallel to changes in the other three processes.

The instituting of specifically economic processes was combined with, and affected by, political processes throughout. In this period, water companies required parliamentary statute to be able to operate, and in that sense, were politically instituted enterprises. Parliamentary lobbying was heightened by extra-parliamentary agitation from the Anti-Water Monopoly Association – a wealthy householder consumer movement, certainly, but one with political leverage articulated additionally through vestry government. The establishment of supply monopoly in 1821 was politically instituted and buttressed by regulation of price and quality. It would be wrong, therefore, to describe the processes of exchange and appropriation as typical commercial markets regulated by the state. Water was not like other commodities in its economic organisation or arrangements, partly because of the material nature of its flows through a singular infrastructure terminating in a final point of delivery: a technological 'monopoly' of distribution. So political processes, in interaction with economic processes resulted in new forms of politico-economically instituted economic configurations. Additionally, a public delivery of sewage services relied on revenues from rates linked to property values, a public economy for dealing with privately generated waste. Significantly, this public economy of waste replaced the earlier market forms linked to cesspit technology, and, in the absence of a tradable product, required a compulsory tax-based flow of revenue.

The great sustainability crisis and the late nineteenth-century water-sewage configuration

The inevitable consequence of the Thames becoming both the dominant source of 'all-purpose' domestic water and the compulsory destination of sewage manifested itself in what was misappropriately called the Great Stink – not that the stench of the river in the summer 1858 was unimportant for its political consequence of suspension of sittings in the House of Commons. But the main sustainability crisis was bacterial pollution of the water supply (Halliday, 2001), especially since the arrival of cholera in England in 1831. It was a liquidity crisis, not a miasmic smell crisis as the medical theorists of the day assumed, including Chadwick, the great reformer behind the next transition. The crisis was not only generated by sewage flowing directly into the source of drinking water, but was compounded by the pollution of groundwater resulting from the continuing breakdown of the cesspit system. Indeed, the discovery of the water-borne and microbial source of cholera by John Snow arose from his detective work on a particular groundwater well in Bow Street in 1854.

If the dynamics of this transition were provoked by an environmental and human biological crisis, the instituting of the new configuration was politico-economic. Many aspects of the first industrial configuration persisted, expanding significantly in scale during the second half of the century. The most dramatic change was the construction of an integrated sewer system, with primary intercept sewers along the purpose-constructed Thames Embankment, fed with networks of underground piped sewers from across London (Hamlin, 1992). It has been described as one of the greatest engineering achievements of the nineteenth century. However, a significant condition for this technological feat was the co-evolution of state finance required for such a vast scheme *and* a development in the form of the local state. To raise the money, a bond was established linked to the assured future revenues from dedicated sewerage rates, and during the same period, a new London-wide organisation, the Metropolitan Board of Works, superseded the fragmented vestry system and Commissioners of Sewers, first with the Metropolitan Sewers Act of 1848, followed by the Metropolis Local Management Bill of 1855 (Hanley, 2006). Critically, the scale of the scheme (it cost £2.4 million at the time) was the outcome of political battles, in particular how far downstream the sewage farm outlets were to be situated. The resultant compromise fell well short of ensuring the restoration of the River Thames. Chadwick and others had had ambitious plans for converting the sewage into manure, to supplement taxation with a revenue stream for the public purse. But despite several attempts the plan never materialised. Instead, Bazalgette, the prime engineer and architect of the sewage system, commissioned the construction of six ships to carry the now separated solid waste to be dumped into the North Sea, while the purified liquid waste was returned to the river Thames. This system operated until 1996, when banned by European legislation. Nonetheless, the new sewage system opened in 1865, after which, unlike most European cities, London never again experienced the cholera epidemics that terrified Londoners in the 1830s–1850s (Hennock, 2000).

The post-crisis configuration therefore instituted a new public economy of waste, which, through its London-wide integration, contrasted sharply with the continuing fragmented private monopolies of water supply (see Figure 2.3). The piped-water system was continuously expanding throughout the second half of the century, embracing the ongoing growth in population. But, as Trentmann and Taylor have shown (Trentmann and Taylor, 2005; Taylor and Trentmann, 2011), inequalities of supply, regular interruptions in frequent periods of drought notably to poorer areas of London and especially the variation and vagaries of pricing water by different companies, all using the institution of rates linked to property values rather than any volumetric basis, resulted in thriving political campaigns. A variety of 'consumer' movements emerged across the country, and included the East London Consumer Defence League and many other Water Consumers' Defence Leagues across the capital. Fuelling the conflict, companies attempted to charge extra percentages on the basic rates if consumers installed baths or WCs, the development of consumer practices destabilising the price institution of the water rate. As far as many water companies were concerned, baths were definitely

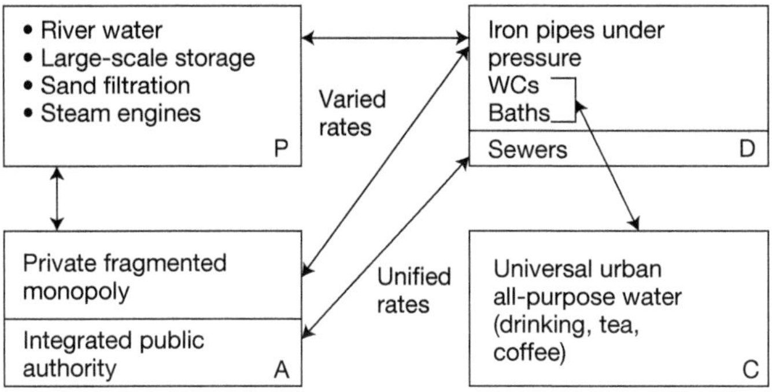

Figure 2.3 The post-crisis, late nineteenth-century water sewerage configuration.

a new 'extra' attracting higher rates, promoting disputes with consumers who considered baths to be normal domestic usage.

Municipalisation and chlorination: the pre-reprivatisation twentieth-century configuration

By 1904, the previous configuration was almost comprehensive with 97 per cent of London's households connected to mains supply and sewage, in contrast to Paris at that time with 17.5 per cent (Goubert, 1989). An even starker contrast between London and Paris existed in sewage systems, where under 20 per cent of towns in France, and fewer households, were connected to sewer systems before the First World War, a contrast that continued for many decades and well into the second half of the twentieth century. It was London that set the pace in Europe for comprehensive coverage of water and sewage infrastructures.

Again, continuities, both technological and economic, can be seen in the transition to the fourth configuration under consideration. There was a developing tension and instability of the previous configuration, in particular between the fragmented and divisive private monopoly sphere of water provision and the integrated sewage public economy. This tension was brought to prominence by consumer movements especially in the face of successive droughts, and led to a significant shift, with the municipalisation of water provision in 1903. The new public economy achieved integration under the Metropolitan Water Board, now responsible for both water and sewers. One of the most significant aspects of the new political economy of water was the standardisation of rates based on property values for both water and sewage across the capital. The municipalisation of water, however, was certainly part of a much wider process of expanding public ownership, and London was relatively late compared with major cities of Liverpool, Leeds and Manchester (Hassan, 1985, 1998; Millward, 2000, 2007; Broich, 2013). Many utilities had come under municipal ownership and control, and across all political parties there was a view that destructive competition over major infrastructures or, conversely, private monopoly power over 'essential' services, was better replaced by a rationalised, publicly accountable, integrated organisation. For many liberals and conservatives, including Neville Chamberlain as Mayor of Birmingham, public ownership did not equate to socialism. There was a widespread concept of a 'natural' monopoly for which public ownership was appropriate. Moreover given the new interdependence between water provision and sewage disposal, with the latter long operating on a non-market basis, at that time public ownership was the only form of organisation that could economically integrate the two under a single system.

Quite contingently, chlorination of water was discovered to prevent bacterial infections in cattle, and the technology was rapidly adapted to the human water supply from 1905 onwards, both in England and the USA. In the USA, the most immediate impact of chlorination was to virtually destroy the burgeoning market economy of bottled spring water for drinking – significant for the problematisation of what constitutes water for drinking (Back et al., 1995). Safe potable public all-purpose tapwater then undermined commercial dedicated bottled drinking water. In the United Kingdom, the process of developing and diffusing the technology of chlorination took almost a further 50 years, with various intermediate stages combining chlorine with ammonia, delivering chlorine through gas rather than liquid, before the fully ready-to-use tapwater became established from the 1950s (Chevalier, 1953). Interestingly for our comparative study, experiments undertaken in the colonial water regime in India led to a modification of the technology, which was then applied in reverse to the United Kingdom water supply. As a bactericide, chlorination was and is a technology of sterilisation, with the dominant aim of making water safe for multiple purposes, of which drinking directly from the tap, was but one. It was a *process of qualification* of water rendering it safe to drink – but clearly not one produced for aesthetic qualities of taste, as it often led to residues of taste and odour.

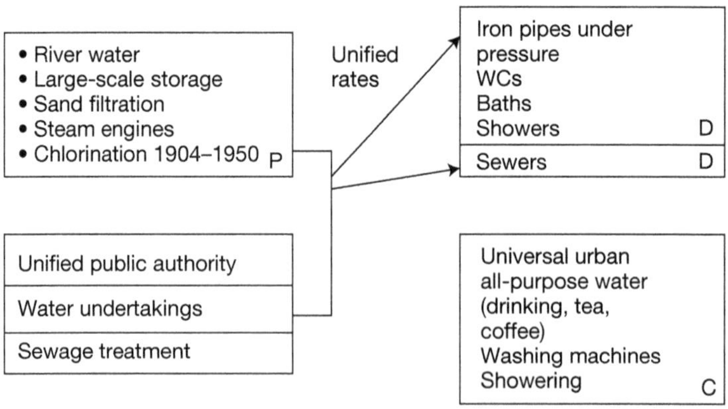

Figure 2.4 The early twentieth-century 'municipal hydraulic model' configuration.

Nationalisation – Regionalisation – Privatisation: 1960–2010

The mid-twentieth century water economy of London can be seen as the consolidation of the 'metropolitan hydraulic model' of a tapwater infrastructure, under public ownership, delivering all-purpose water to households on a 24/7 basis. However, this was not the culmination of history, a stable or sustainable end-point. The subsequent next half century from the 1960s witnessed quite radical and continuous configurational change. As in previous transitions, there was again a complex and interdependent mix of dynamics, comprising infrastructural renewal and development, scale constraints and subsequent upscaling, failing financial flows, and environmental unsustainability in terms of both abstraction and pollution. As in previous transitions, politics and political conflicts were inextricably linked with each element of the mix.

This latest period divides into two phases: the first characterised by state rationalisation of water and sewage provisioning systems, seen by some as a formative prelude to the second phase of privatisation (Bakker, 2003a). In the post-war period of economic growth, demand for water grew rapidly, especially by industrial but also by end consumers, as slum clearances and new housing development coincided with the final stages of the domestic 'internalisation' of water supply and toilets (Hassan, 1998). Increasing demand intensified the environmental threats of both over-abstraction and unregulated pollution, forming the background to an upscaling of economic organisation, and, at the same time, the emergence of 'nationalised' water resources. Two major bits of legislation, both initiated by Conservative but implemented under Labour governments, the Water Resources Act of 1963 and the Water Act of 1973, brought about a two-stage transformation which has been deemed 'revolutionary' (Hassan, 1998; Parker and Sewell, 1988).

Although there had been integration across London of water and sewage provision, there was still a general fragmentation of control and governance over its upstream water resources, a pattern shared across the country in terms of water supplies. Major cities such as London continued to dump sewage sludge directly into the sea, and sewage treatment was even more fragmented than water provisioning. A severe drought in 1959, followed by serious flooding the following year, exposed the fault-lines in the integration and scale of water and sewage undertakings. The 1963 Water Resources Act was the first move towards scaling up governance of water resources, providing for legal recognition of ground and surface water as a national public economic good. It established 29 River Authorities which covered whole river basin catchment areas, giving them the power to license any abstraction of national ground or surface water for the first time. A national Water Resources Board was created for oversight. In effect, the Act instituted new national property rights over the country's water resources – an historical innovation of major significance.

However, the 1963 Act failed to correspondingly scale up water provisioning undertakings, still largely under municipal control, or coordinate them with the even more fragmented sewage treatment organisations. There continued to be over 1000 separate water undertakings, and over 1500 sewage treatment works, with no powers of coordination or governance over their operations at the River Authority spatial scale. With expanding demand for water, these disjunctures of scale and coordination became more acute particularly because of increasing reliance on low-level, 'dirty', surface resources as against upland 'cleaner' sources. London, almost alone of major cities, relied on the Thames river catchment area, lacking any high-level dam-plus-reservoir 'clean' supplies. Sixty percent of sewage undertakings failed to meet Royal Commission standards by the late 1960s.

Nonetheless, the 1963 Act pre-figured the 1973 Water Act which further consolidated river basins as the integrated scale for both water and sewage, now creating ten Regional Water Authorities, effectively de-municipalising the control over water. The Metropolitan Water Board was abolished and replaced by the Thames Water Authority now responsible for the whole catchment area and all towns and villages within it. The disjunctures and subsequent consolidation of Integrated River Basin Management also addressed the financial 'leakages' of the municipal hydraulic model. For much of the post-war period, and increasingly during the 1960s, water rates flowing to municipal governments ran substantial deficits, and had to be supplemented by central government taxation. Although implemented by a Labour government in 1974, the legislation had a strongly Conservative stamp, the new RWAs being obliged to become self-financing, including through the gradual introduction of water meters. This has been viewed as a fundamental shift in the understanding of water as a public good: 'The Act moved the industry towards the philosophy of water as an economic good rather than as a subsidized public health service' (Parker and Sewell, 1988, 767). However, from an IEP perspective, British water provisioning was never like the National Health Service, more closely resembled by the Irish system. This shift was certainly fundamental in terms of a move towards full cost recovery, but remained distant from market commodity pricing, as indeed water pricing does to this day, even post-privatisation.

The de-municipalisation and scale expansion unsettled pre-existing linkages between monetary and water flows: how water provisioning was financed and paid for. In the long historical trajectory just surveyed, payments of water and sewerage rates calibrated on property values were the form of payment initially established for the urban wealthy, the ones connected to pipes. In the United Kingdom, it was never a service free at the point of delivery. On its historical introduction, as an instituted form of exchange, it was therefore certainly not designed as a redistributive form of water taxation. After municipalisation, there was an alignment between water rates and local government taxation, both based on the principle of proportionality to property values. Whether for refuse collection, road maintenance or sewage and water services, there was a similar system of payment through billing for 'rates'.

However, throughout the 1960s and until the 1973 Act, rate-based financing failed to maintain or renew infrastructures, particularly sewage, and was supported by the rate support grant from central to local government. During this period, over half of the sewage works failed to meet legal standards, and by 1973 'the industry's underground assets were on the point of collapse' (Hassan, 1998, 111). The institutional channels established for financing both water and sewage provisioning continued to be a political battlefield. The Labour government introduced a Water Charges Equalisation Act, attempting to overcome the huge regional differences in charges for water and sewage, with the South West in particular being highly charged. The Thatcher government repealed this law, and at the same time removed the rate support grant to supplement water rates, leading to a 20 per cent hike in charges, with disparities between regions increasing even further. Under Labour from 1974, following the fiscal crisis and the intervention of the International Monetary Fund (1977–78), and then under the Thatcher government from 1979, the water and sewage industry was starved of capital investment. Just as during the period following the 2007–8 financial crisis, reducing the Public Sector Borrowing Requirement (PSBR) became an imperative (Parker and Sewell, 1988, 768–70). Both capital and income flows for maintaining and developing public infrastructures were in such crisis that for the industry as a whole 'the main problem was the damaging and arbitrary effects of the government's financial policies' (Hassan, 1998, 161). Thus, the destabilisation of financial flows combined with political conflicts set the scene for privatisation.

Whether by political design, internal incoherencies embedded in earlier reforms, or both, the failure to develop public systems of provision and financing at the new geographic scales opened the gateway to privatisation, and a transition to a further reconfiguration of London's economy of water. The IEP analysis of this contemporary phase of economies of water aims only to highlight some of its key aspects, relying heavily on much more comprehensive accounts (Bakker, 2003a, b; 2010; Hassan, 1998). There is an historical irony that this 'long-duration' story almost begins and ends with state-instituted private monopolies of water. The ten Regional Water Authorities were sold to private companies in 1989, but under a regulatory regime and architecture unique to water as an industry. As Bakker has insisted, there was no water market and water did not become a commodity (Bakker, 2003a, 74–5), evading any neat categorisation and challenging any state-market dichotomy. This reconfiguration by privatisation was also characterised by a 'shifting place of the water economy in nature', partly because it was critically shaped by a succession of European Directives requiring major infrastructural investments in sewage systems and pollution control under a mandatory timetable (Helm and Rajah, 1994). These Directives required formal commitments to capital expenditure on underground infrastructures to meet enhanced environmental standards.

Water privatisation was one, and not the first, of a succession of privatisations of public utilities undertaken by the third Thatcher government, all of which instituted particular forms of market or quasi-market forms of economic organisation. However, water privatisation was unique in instituting a non-market state-guaranteed monopoly organisation, with separate regulatory powers for controlling a financial regime (prices charged to customers and capital investment programmes), drinking water quality and environmental protection of water resources (abstraction and pollution). In characterising the new economy of water configuration, it is worth considering briefly each of these three of its dimensions in turn.

The regulatory body responsible for pricing and capital investment, Ofwat, has been accorded increasing powers to control and cap prices charged to consumers, using a price formula which included commitments to levels of capital investment in infrastructure (the RPI+K formula, also unique to privatised water companies) (Helm and Rajah, 1994). In effect, this provided guaranteed rates of return on capital investment, resulting in a rapid rise in share values, post-privatisation. State monopoly and pricing systems afforded the water companies unique opportunities in capital markets, manifested in a more than doubling of capital investments in the six years following privatisation. For London and Thames Water, the physical manifestation of this investment has parallels with the mid-nineteenth century engineering feats of Bazalgette: first a £250 million Thames Water Ring Main, with 80 kilometres of potable water pipework and eleven underground pumping stations (Fullalove, 1994); second, and in its early stages of development, a new super-sewer, the Thames Tideway Tunnel, due to start construction in 2016, and designed to minimise the level of sewage discharges into the river basin.

The pricing of water to consumers under the Ofwat regime has been under continuing political pressure and regulatory reform. The Labour government proscribed water disconnection for non-payment of bills in the 1999 Water Industry Act, previously permitted by Thatcherite legislation, so effectively securing a non-market right to water. A windfall tax in 1999 was followed by a periodic review by Ofwat that reduced the real-term price of water by 12 per cent up to 2005. The requirement for water companies to universalise charging by meters was quietly dropped, resulting in huge disparities between different regions, with London under Thames Water having amongst the lowest level of metering at 26 per cent compared with Anglian Water at 61 per cent in 2008–9, along with the lowest average annual water bills (£300 as against £480 for the South West) (Walker, 2009). Regional private monopolies are relatively insulated from any national-scale market pressures. Although 'river basin' privatisation has entrenched a scale of economic organisation exhibiting huge regional variations in pricing and investments, the monetary valuation of water to consumers remains, by and large, especially when including sewage services and standing charges, closer to a tax than a price. However, although in appearance there seems little to distinguish water bills before and after privatisation, the key change is that consumers now contribute to the profit margins and share values of private corporations. State price regulation guarantees that flow of cash from consumers to shareholders. In this perspective, it is not surprising that Thames Water is now owned by an Australian bank operating as a financial vehicle for institutional investors, including pension funds, given that the state underwrites share value and dividends, and ultimately assumes responsibility that water companies are too necessary to fail.

The Water Industry Act of 1991 for the first time made supply of non-potable water to consumers a criminal offence, and the quality of water was placed under the authority of the Drinking Water Inspectorate. The regime of qualification of domestic water supply to the public is still dominated by the characteristic of safe water, necessary to 'protect public health' in a way that inspires trust and confidence (Drinking Water Inspectorate, 2012, 2013). To that end, a rigorous regime of inspection and detailed geographical annual reporting of micro-biological and chemical risks was introduced, with over 100 possible contaminants measured from source to tap. Supplementary to this objective,

> Consumers expect their drinking water to be clear and bright in appearance and free from discernible taste or odour. In recognition of this the regulations stipulate national standards for colour, odour and taste. Companies are required to investigate the cause of any adverse result.
>
> (Drinking Water Inspectorate, 2013)

The DWI annual reports thus document 'incidents' where consumers complain of excessive chlorine odour, turbidity or discolouration. Apart from quite striking contrasts in softness and hardness of water in different regions, it is nonetheless clear that there is some regional variation in levels of residual chlorine, and reports occasionally refer to inadequate practices of imprecise manual dosing of reservoirs with chlorine. Yet, the United Kingdom generally has levels of residual chlorine well below legal limits. There is, relatively speaking, a high level of quality standardisation. Domestic tapwater supplies therefore have a quite distinctive quality regime, contrasting sharply with the quality regimes for bottled water (discussed in Chapter 3 below). In contrast to bottled water supplied by private companies where the same brand of water has the same quality of water wherever it is bought, the Drinking Water Inspectorate at most attempts to ensure that all companies supply water of similar quality to all consumers. Similar within certain parameters but not the same, as annual DWI reports demonstrate, there remain significant quality variations between water regions and between different water companies. As we will see in Chapter 3, these differences, and this tapwater quality regime, provide a significant context for the emergence of bottled water markets.

Nonetheless, consumers now take this relative standardisation so much for granted that it is important to step back and view it anthropologically in order to recognise the extraordinariness of a regulatory regime that imposes public quality standards on a final consumption good produced by private multinational corporations. Above all, the quality of tapwater is constantly evolving. The level and process of chlorination, as already noted, developed over 50 years, and even now is subject to new standards. One of the most striking recent examples of developing qualities of tapwater emerged in relation to the progressive elimination of lead in drinking water. As late as the 1980s and early 1990s, both UK and European Regulation imposed new limits on lead in drinking water. That required the complete overhaul and replacement of domestic plumbing infrastructures, first introduced in the early eighteenth century with the New River Company (Boyd et al., 2001). The estimated cost of lead plumbing replacement was up to £8 billion, and much of the responsibility lay with the domestic householder, although the communication pipe from the mains to the private house property boundary was also often in lead, and the responsibility of the network, public or private (Millstone and Russell, 1995; Hayes, et al., 1997). Quality, property and financial responsibility were all institutionally 'connected' in this extensive upgrading of domestic water infrastructures. (The quality of gas or of electricity are, of course, also strictly regulated to ensure safety and standardisation, but they are not final consumption goods.) The historically ongoing qualification process of tapwater is therefore quite distinctive.[3] It is as if all firework companies not only had to produce safe fireworks, but also fireworks so similar that one can scarcely tell the difference between what one company produces from another. Despite privatisation, the Drinking Water Inspectorate regulatory regime 'brands' tapwater with the economic quality of a public good: universal, standardised protection of public wellbeing and safety.

The final dimension of the regulatory control of the privatised economy of water for Thames Water and companies responsible for other river basins, is environmental. Much of the impetus for regulatory control of the industry stems from the European Union, and in particular the Water Framework Directive, passed in 2000 (EU, 2000). This Directive obliges member states to maintain or restore river basin surface water to a defined Good Ecological Status by 2015 (Acreman et al., 2008). In the United Kingdom regulatory framework, the Environment Agency is primarily responsible for the monitoring and enforcement of the requirements of the Directive. Although in the early days following privatisation, the Environment Agency proved subordinate to the economic interests of Thames Water, notably in failing to prevent abstraction of water from the River Kennet in the Axford case (Bakker, 2003a; Bateman et al., 2000), substantial investments have been enforced to improve environmental quality and management of river basins. River water quality is now as high as it was since before the Industrial Revolution, and is improving (DEFRA, 2010). The major commitment of the Thames Tideway Tunnel, limiting sewage discharges from flooding, is a £4 billion capital investment directly arising from environmental regulation. The United Kingdom has travelled some distance from being 'the dirty man of Europe', with Bazalgette's barges dumping London's sewage in the North Sea (Rose, 1991). The Environment Agency's responsibility for controlling national public water resources, by licensing abstraction and controlling pollution, must be seen as an integral complement to the two other members of the regulatory triumvirate: Ofwat and the Drinking Water Inspectorate.

The state instituting of private, river-basin scale monopolies operating under financial, quality and environmental regulatory controls forms a highly distinctive economy of water configuration, represented in Figure 2.5 below. The private monopoly economic agents, such as Thames Water, were co-instituted and have since co-evolved, with the regulatory triumvirate and

3 Although distinctive as a final consumption good, every householder in the UK, and no doubt elsewhere, will be aware that the qualities of gas and electricity services are also constantly evolving with new regulations requiring upgrading of domestic infrastructures at the expense of the consumer.

their respective powers of control. From an IEP perspective, the economy of water is constituted by a configuration with multiple actors, corporate and regulatory, and their relative powers of control over processes of production (qualification), distribution, exchange and consumption. The power–relational configuration as a whole marks out this economy as distinctive, both conditioned by its prior trajectory – most conspicuously in the pricing institution of the property-based water rate – and conditioning future crises and further transitions. Droughts and floods, climate change and population concentrations will no doubt place further stresses on the river-basin scale of organisations now enshrined in privatisation monopoly boundaries.

So, a final supplementary element to this tapwater configuration, to be explored more thoroughly in the next chapter, involves the development of a dedicated bottled water market. Arriving much later in the United Kingdom than many European countries, it is still generally on a much smaller scale than its continental counterparts (Euromonitor, 2009). In configurational terms, from the long durée historical perspective, bottled water can be seen as the most purely market water to emerge, with major international companies operating in a competitive market. As with most markets, it is also highly regulated, and in Europe strict distinctions are made between spring and mineral water, with prescriptions on quality and

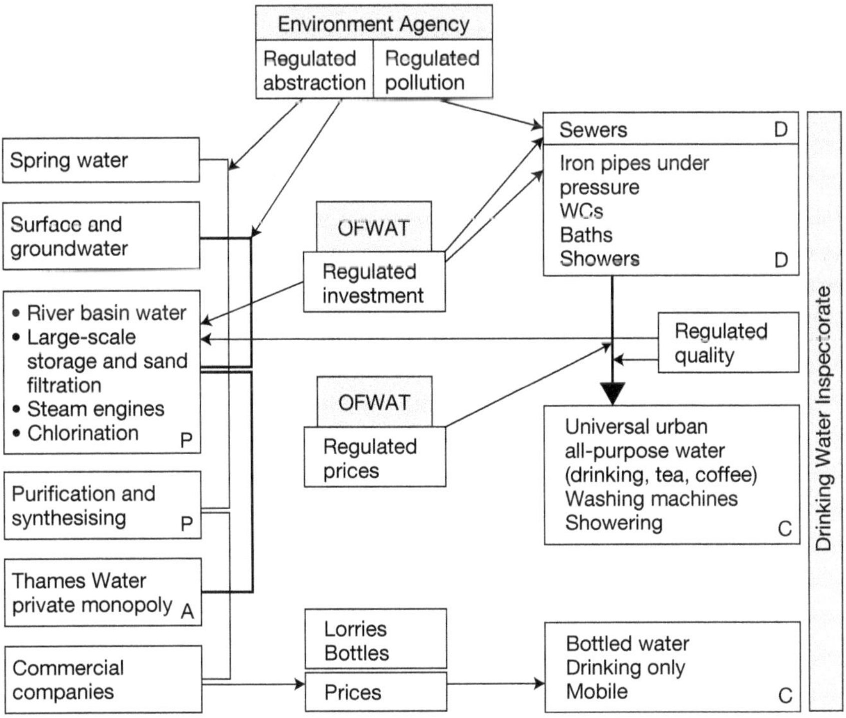

Figure 2.5 The instituted private monopoly economy of water.

process. Nonetheless, the irony remains that bottled water, as a purely commercial product, is licensed by public authorities to extract 'public' groundwater, appropriate it, and market it with all the blandishments of branding. Carrying around bottles of water and drinking water on the move have become new consumer practices. Bottled water, especially in the UK, represents a tiny fraction of water consumed as a beverage, even of tapwater drunk in its various forms (raw, boiled for tea, coffee, etc.). It constitutes a semi-autonomous additional system of appropriation, distribution, exchange and consumption, going through supermarket systems of regional distribution centres, which transport spring water by lorry from Scotland or Wales to South East England – as well as from France and Italy and further afield. As we shall see, bottled water has its own very distinctive qualification processes, varying sharply between the UK, Germany and Italy. With distinctive modes of qualification and marketing, it displays an extravagant range of prices, qualities and brands, with a multiple of 20 to 30 times from cheapest to most expensive per volume, characteristic of market differentiation. In configurational terms, however, and in ways that will become clearer by comparison, bottled water finds a particular and distinctive place only within the overall configuration, and is best understood in terms of this relationality. So, in this final configuration, bottled water finds a small and separate place.

Endogenous transitions

In this chapter, a long historical view has been constructed in order to examine successive transitions of economies of water. It has been a tortuous history, marked by crises and upheavals, and no linear developmental trajectory. Unquestionably, urbanisation and industrialisation, including adoption of technologies such as steam power and iron-casting of pipes, produced a major rupture from pre-industrial periods. Yet, even there we can see the industrial system pre-figured in the wooden pipe system of the New River Company and other water companies, as well as the remarkable endurance of the property-based water rate system, surviving even the recent introduction of metering for private domestic units.

Certain key points emerge from the comparison of successive transitions. Firstly, transitions are always transformations of previous configurations, with elements of continuity, modified by the new relational dynamics of new configurations. There is co-emergence, co-evolution of new production, distribution, appropriation *and* consumption processes. In this analysis, therefore, it makes more sense to understand the emergence of new consumption practices always as both transformations of previous practices, and in the relational context of a given historical configuration. In this respect, consumption work, work necessary for consumption after the point of final delivery and appropriation, is integral both to consumption practices themselves, and to the division of labour upstream of final delivery. Domestic water storage and treatment (filtering, boiling, alcoholisation, flavouring, etc.), domestic distribution through plumbing systems, disposal and treatment of waste water, and practices of defecation,[4] are integral to the nature of the water delivered, and systems of production, exchange and distribution. New processes of preparation for final consumption, and novel coordination with other consumption activities – e.g. carrying bottled water for sports training – are integral to the emergence of new consumption practices.

Secondly, the comparison of London's successive transitions demonstrates both persistence and variation of some of the economic dynamics, in interaction with both political initiatives and responses. Long-term expansion of urban population and the dynamically growing demand for ever higher standards of water quality and domestic convenience (washing machines, showers, etc.) are examples of the progressive ratcheting up of standards (Shove, 2003). London boasted the purest water both in 1840 through its regulated quality of reservoirs and slow-sand filters, and again in 1905, with the additional treatment of chlorination. The elimination of cholera and typhoid were major achievements in progressive improvements of water quality. And today, Thames Water boasts some of the highest quality water in Europe.

Thirdly, there were distinctive contrasts between the different transitions: no one type of transition. The sociogenic sustainability crisis of the mid-nineteenth century – the Great Stink – stands out as resulting from the first industrial configuration's cast iron pipes, WCs, *and* legal regulation, then provoking transition to the second industrial configuration. Subsequently, the tensions between private monopoly water companies in London and public sewerage systems in the late nineteenth century again exemplify how a previous configuration provides the dynamic context of its own demise and replacement, in this case by a fully integrated, municipalised water-sewage organisation, in part driven by consumer movements, as well as wider political strategies of economic rationalisation. Significant price differentials between one borough and another and irregular supply especially during a succession of droughts, generated socio-economic and environmental tensions to induce further change. Finally, from 1963 to 1989, the extended transition from the post-war 'municipal hydraulic model' to the contemporary instituted private monopoly was driven by tensions between financial and water flows, an internal breakdown of the municipal model, combined with an upscaling to meet the challenges of sustainable supply and pollution control. The river-basin scale, dividing the country into ten water authorities, de-coupled water governance from the municipal, or indeed any other, established governmental scale. Yet, this scale disjuncture certainly did not arise within any long-term vision for subsequent privatisation. Indeed, the establishment of the River Water Authorities has been characterised as a nationalisation,

4 In drawing attention to a contrast between England and France, squat toilets persisted in the latter until the late twentieth century, in contrast to the established domination of the seated WC in the UK, and there remain many distinctive designs of toilets conditioning the 'embodied practice' of defecating and urinating. A converse contrast lies in the diffusion of bidets in France, and, until recently, their virtual absence in the United Kingdom.

superseding municipalisation, of public water provision, albeit by a Conservative government. A given configuration expanded, driven often by economic and commercial growth, only to develop contradictions and tensions that paved the way for further transformation. The dynamics of a given configuration constitute immanent and emergent causalities within that configuration, as well as generating new interactions with the environment, political processes and culture, which in turn contribute to configurational transformation.

Fourthly, water developed, changed its characteristics, as both a private and a public good. The evolution of water as an economic good can be seen as contributing to the co-evolution of the differentiation and interdependence between 'the private-market', 'private-domestic', and public goods, a reconstitution of both the 'private' and 'public'. This was, throughout, an intensely political process, and in that respect, the successive transitions demonstrate the interaction between different processes of instituting economic organisation, both private and public. The Bazalgette technological sewer system co-evolved with the development of both the political organisation of the capital and the organisation of the public economy. Political tensions and conflicts represent a continuing strand through the successive transitions, lobbying behind the monopolist water companies countered by the lobbying of anti-monopoly property interests, and later, consumer defence associations. A dynamic mix of the instituting of novel commercial economic organisation, politically instituted market organisation, market regulation of price and quality, as well as the instituting of radically new forms of state-underwritten private monopolies, are testimony to the complexity of transitions from one configuration to the next. In the contemporary period, it has been argued that, through regulation by the Drinking Water Inspectorate, tapwater has all the appearance of a public good, even when produced by profit-making, multinational enterprises. In short, it is not only a question of getting away from static dichotomies between 'public' and 'private', state and market, but understanding that the institutionalisation of 'public' or 'market' of contemporary tap or bottled water have co-evolved from earlier forms, across many phases, going back to the eighteenth century and earlier.

Fifth, and finally, water is not only distinctive as an economic good, but also as a material entity, environmentally and biologically, in ways that conditioned the interactions between polity and economy. The sources and geographies of water are not only the continuing lifeline of urban development in this narrative, but also the bearers of bacteria and viruses pathogenic to human biology. However much humans were responsible for pollution and the spread of diseases, the diseases were responsible for the deaths. Water presents particular challenges to distribution and purification, but also, as the different transitions demonstrate, continuing tensions and conflicts around appropriation: who, what and how does any social organisation 'own', 'control' a river, an aquifer or rainfall (my water in my water butt?), as they are naturally given. In that respect, the natural properties of various natural waters, independent of human constructions, continuously condition, constrain and stimulate those water constructions: political, economic or cultural (including epistemic). The transitions analysed above are of London, in its physical dependence on the Thames, the Lea, and other rivers; on the particular aquifers and rainfalls; on the floods and droughts; and on the organisms that thrive in this environment. Apart from economies of scale arising from a large urban conurbation, London contrasts with many other cities in the United Kingdom in having sufficient resources from its immediate river systems, without relying on dams and long-distance transfers. This geographical privilege no doubt contributes to Londoners' low water bills.

In tracing the successive transitions, perhaps one of the most striking features is just how long the United Kingdom's capital took to develop a relatively comprehensive, physically and economically uniform, economy of domestic tapwater. Moreover, it would be foolhardy to think that this process of economic reconfiguration and scaling has reached its final and enduring phase, immune from future upheavals, political, economic or ecological. When, in future chapters, we explore some very different transition processes in the radically contrasting environments of Delhi, Mexico City and Taiwan, with their own distinctive nature–economy–polity interactions, it will enable us to see the specificity of the UK history, by stepping outside it. Not least, the transitions analysed above have all been endogenous, arising from within a given historical English configuration, although latterly with a strong impetus from European regulatory developments. Elsewhere, colonial occupations and foreign invasions introduce radical breaks, with major consequences for the trajectories of developing economies of water. But before going further afield to more extreme contrasts, we turn first to a comparison within Europe, where bottled water is also a complement to universally accessible tapwater, in order to better understand the drinking of bottled water in the UK.

3 Configuring bottled water in Europe[1]

Adrian Evans and Mark Harvey

Introduction

In this chapter we shift attention away from a predominant concern with the provision of tapwater to concentrate on the formation of the market for bottled waters in three European countries (the UK, Italy and Germany). However, as will become clear throughout the chapter, we never leave tapwater far behind, as it forms an essential context for understanding the different characteristics of the markets for bottled water in these countries. For example, it would not be possible to understand the demand for bottled water in certain regions of Italy without considering the lack of trust in the safety of local municipal supplies of tapwater.[2] Similarly, it would not be possible to understand certain aspects of the demand for bottled water in the UK, without considering how chlorination as a water treatment method impacts upon consumer perceptions of water quality and taste, or how regional variability of the taste and quality of tapwater contrasts with the consistency and uniform quality of bottled water taste for a given brand.

Whilst the provision of tapwater provides an essential background to understanding the formation of the market for bottled waters in these countries, it is by no means the only factor, or point of comparison, which needs to be taken into account. Indeed, it would be a mistake to only compare bottled water with tapwater without considering the full range of alternative drinks that are available on the market. For example, in the UK, bottled water accounts for only 6 per cent of the market for non-alcoholic drinks compared with 40 per cent for soft drinks (Living Costs and Food Survey, 2009). Furthermore, many large soft drinks companies in the UK also produce bottled waters, so the two products are deeply

1 We would like thank Mara Miele and Steffen Boehm for their help, advice and support in organising the Italian and German bottled water case studies. We would like to thank Carlotta Lami for carrying out the Italian diary research. We are exceptionally grateful for the contribution of several key informant interviewees in the UK, Italy and Germany, who generously gave their time to contribute to the research.

2 This lack of trust in the safety of municipal supplies has arisen both due to past agricultural pollution events and due to the natural geo-hydrological characteristics of certain Italian watersheds, which can lead to tapwaters with higher than permitted levels of arsenic.

interconnected within their overall market strategies. It is also clear that the consumption of soft drinks, cordials, teas and coffees can have significant environmental and health impacts. The key comparison is between how people consume water in multiple ways in different societal contexts. It will be argued that the appropriate comparison for bottled water is not with tapwater in isolation but of its place within the overall pattern of consumption of bottled beverages and tapwater. It will be shown that bottled water occupies very different places in these overall patterns in the UK, Germany and Italy. Moreover, in subsequent chapters, a broader contrast emerges between the European context where bottled water is at most complementary to, rather than a substitute for, tapwater, and societal contexts lacking potable tapwater where bottled water may be the principal conduit for water fit for drinking.

In both popular and academic literatures concerning bottled water and especially in the campaign literatures of certain environmental NGOs a frequent argument is made that drinking bottled water represents an act of conspicuous consumption, as socially unacceptable as smoking whilst pregnant according to some.[3] It is suggested that an almost identical product can be sourced for a vastly reduced price and at far less cost to the environment by simply turning on the tap. Indeed many commentators have questioned the very existence of bottled water markets in countries which have safe and almost free drinkable tapwater supplies (Gleick, 2010; Royte, 2011; Barlow and Clarke, 2002; Salzman, 2014). Gleick (2010) in particular is at a loss to explain why consumers would want to purchase bottled water as it is so similar, yet so costly and environmentally damaging. The only explanation he offers is that consumers have been duped into purchasing this product by the marketing machines of bottled water companies. In other words, he argues that the demand for bottled water has been manufactured.

In contrast to Gleick (2010) we want to paint a different picture of the demand for bottled water, placing it in the overall configurations of how bottled water is produced, marketed, distributed and consumed in different contemporary European societies. Rather than viewing the market for bottled water as simply or solely a case of manufactured demand, such an approach would consider the complex ways in which bottled water markets emerge and develop. For example, as we shall see, elaborating such an account for the UK will force us to consider diverse yet interconnected issues such as the use of chlorine in tapwater treatment, the dominance of the large multiples within the UK grocery sector, the role of European legislation in defining bottled water categories, and the interconnections between soft drink and bottled water producers. The question is why mostly better-off consumers in the UK purchase bottled water, and in much smaller quantities, than the generality of German or Italian consumers. Societal comparison enables us to address head-on the issue of whether bottled water is just a substitute, at a much higher price, for tapwater. It does so because we are comparing firstly

3 See for example, http://storyofstuff.org/movies/story-of-bottled-water/ [Access date 13 August 2014].

different tapwater contexts, secondly different ranges of qualities of bottled water, and thirdly different overall markets for bottled beverages. A benefit of this approach is that it enables us to bring together processes of water quality differentiation (tap and bottled) with prices in market exchanges (Callon et al., 2002; Harvey et al., 2004; Heuts and Mol 2013).

The remaining sections of this chapter develop this IEP approach to understanding European bottled water markets in more detail. Firstly, the methodologies used to explore the different European bottled water economies are outlined. Secondly, the three comparative national case studies of bottled water markets are each systematically analysed in terms of configurations of production, distribution, exchange and consumption in the UK, Italy and Germany, although in much more depth for the UK. These countries were selected to represent some of the diversity in bottled water markets across Europe, with the UK representing a small but growing market in Northern Europe; Italy representing a large and established market in Southern Europe; and Germany representing a large, Central European style market, with a preference for highly mineralised sparkling mineral waters. Of course other European countries would have made equally interesting case studies. However, the point of a comparative IEP approach is not to provide an exhaustive account of the nature of bottled water markets throughout Europe, or even, for that matter, to produce a definitive account of bottled water markets within our case study countries. Rather, the aim is to analyse what is driving variation in economies of bottled water in different societal settings. The chapter concludes by arguing that a relational and configurational approach is critical for explaining why consumers drink water in markedly different ways in different societies. In particular, we argue that the issue of sustainability in relation to bottled water needs to be placed in a more holistic account of overall social consumption patterns of how water is drunk, as plain tapwater, tea, coffee, bottled water and soft drinks.

Methodology: tracing bottled water from 'source to sip'

Adopting an IEP approach to bottled water markets involves addressing bottled water from 'source to sip'. Unlike traditional supply chain approaches which tend to privilege vertical linkages in the supply chain (Glennie and Thrift, 1993), an IEP approach is equally concerned with more 'horizontal' relations between, for example, water-consumption processes and other everyday practices, or between production processes, geo-hydrological contexts, and legal or regulatory environments. In order to address bottled water *production*, a range of dimensions are explored: techniques of bottled water production; the connection between production processes and environmental/geo-hydrological factors; the number, size and degree of concentration of producers in a given market; the relationships between producers and retailers; the strategies water producers use in manufacturing their whole range of bottled beverages. In order to compare societal *distribution and retail* systems we explore the following: the power relations between retailers and manufacturers and the significance of supermarket 'private labels'; the types of outlet in which bottled waters are available and the differences between outlets; the varieties of bottled water available in different stores; and the classifications, qualifications, marketing and pricing of different bottled water products. In relation to processes of bottled water *consumption*, the analysis highlights social, spatial and temporal variations in bottled water consumption; how the consumption of bottled water relates to the consumption of other drinks (e.g. tapwater, soft drinks, hot drinks etc.); how the consumption of bottled water relates to a range of everyday activities (such as eating, working, travelling, exercising etc.); and whether consumers need to undertake further 'consumption work' (Glucksmann, 2013) before the product is ready to consume. Finally, and most crucially, the analysis pulls these different dimensions together to understand how they form different *configurations*, different overall economies of bottled water, in different contexts.

In order to achieve this complex analysis, a multi-method approach has been adopted, drawing on a variety of sources of primary and secondary data relating to European bottled water markets. Secondary data from market reports gave an overall picture of the size and structure of bottled water markets (e.g. Euromonitor). A range of quantitative techniques (such as factor analysis) was used in the UK to examine socio-demographic, age, and seasonal patterns of bottled water consumption, analysing Living Costs and Food Survey (LCF) data. Market audits were conducted in the UK, Germany and Italy to explore national and local variations in the availability and prices of different bottled waters. We interviewed over 20 key informants from bottled water and soft drinks manufacturers, regulators, environmental NGOs, consultants, trade organisations and retailers. Finally, we undertook qualitative diary research with consumers in the UK and Italy to explore everyday drinking practices and the spatial and temporal rhythms and routines of drink consumption. Small samples of consumers, with a range of different age groups, genders and socio-economic backgrounds, were asked to record each instance of drinking water, in whatever form, on three days of a given week. Over the coming sections we draw on all these resources to present an analysis of variations in the configurations of bottled water markets in the UK, Italy and Germany.

Situating the UK bottled water economy

Compared with many other European markets the UK market for bottled water is relatively young and still experiencing a period of growth. It is also far smaller than the market for bottled water in certain other European countries. For example, consumers in Italy and Germany drink almost four times as much bottled water per capita at 0.4 and 0.35 litres per person per day respectively compared with a mere 0.1 litres in the UK (Euromonitor, 2009). The UK market for bottled water is also relatively small compared with the UK market for soft drinks, in contrast to Germany, where bottled water dominates the market share for non-alcoholic drinks, with soft drinks well behind.

So first, let us situate bottled water in the overall bottled beverages market. Expenditure on bottled water made up only 6 per cent of the total expenditure on non-alcoholic drinks in 2009 in the UK, compared with 40 per cent for soft drinks (Living Costs and Food Survey, 2009). In the light of this basic market data one might ask why so much attention has been devoted to bottled water in the UK as an environmentally damaging product (Wilk, 2006; Hawkins, 2009). It is known that, in contrast to bottled spring or mineral water, every litre of soft drinks uses two to three times more water in production. Indeed, Coca-Cola even boasts in its advertising that it has reduced the amount of water it uses per litre of Coca-Cola from 2.7 litres to 2.08 litres between 2004 and 2012.[4] Equally the health impacts of the much larger soft drinks market have been widely discussed, while bottled water has no such ill effects, even some possibly beneficial ones (French et al., 2003; James and Kerr, 2005). In terms of environmental impact, soft drinks as a way of consuming water are both more problematic and, in the UK, are purchased on a much greater scale than the environmentally more highly stigmatised bottled water. Of course, the environmental impact of all that is eaten or drunk must be a matter of concern, but it is questionable whether bottled water should receive any special attention because of its superficial substitutability with tapwater.

In contrast to the markets in Germany and to a lesser extent Italy, the vast amount of bottled water consumed in the UK is still rather than sparkling water and indeed still water accounted for 1344 million litres or almost 60 per cent of the entire market in 2008 (Euromonitor, 2009). However, as one of our key informants indicated, this has not always been the case:

> When mineral water first came onto the market the sparkling mineral water outsold still by quite a substantial amount and I think that was because people said, well sparkling I can't get that out of the tap, it is slightly different, it's half way to a fizzy drink ... but as time went on and went through generations, I think, I don't know what the number is now, but I suspect that still outsells sparkling 2 to 1, 3 to 1.
>
> (Retired CEO of a large soft drinks company)

There is also a relatively large market for flavoured and functional waters in the UK (569 million litres in 2008, Euromonitor, 2009), which can in part be explained by the fact that many soft drinks producers also sell bottled water and hence cross-over products such as flavoured and functional waters are relatively simple and convenient to produce.

Finally, in terms of the nature of the product itself, the vast majority of bottled waters sold in the UK are bottled from natural sources (i.e. they are either legally classified as 'Natural Mineral Waters' or 'Spring Waters'), rather than being processed waters. As one leading consultant for the bottled water industry informed us:

4 http://www.coca-colacompany.com/setting-a-new-goal-for-water-efficiency [access date 13 August 2014].

> Europe's not really the place for processed waters, if you look at what we have here it's: 'I want it natural, I want it pure, I want it safe'. In the States it's: 'I want it safe and I wouldn't mind it natural and pure'.
>
> (Consultant to the bottled water industry)

As will be seen later, the salutary tale of Coca-Cola's unsuccessful attempt to introduce a processed water, Dasani, into the UK market reinforces this point.

This initial market overview provides a first insight into the place of bottled water in the beverage marketplace in the UK. The strong and growing presence of flavoured and functional waters hints at the role of soft drinks manufacturers in shaping the market. The decline in market share of sparkling water compared with still water suggests a changing role and usage of bottled water. Finally, the lack of a large market for processed waters tells us something about both the preferences and ideologies of UK consumers and about the legal and commercial organisation of the UK market. Processes of qualification (sourcing, refining, packaging, branding) create a sharper distinction between spring water and tapwater than between processed water and tapwater. Over the next few sections we will build on these insights to develop a more in-depth analysis of the fundamental economic processes of consumption, production, distribution and ownership/appropriation that interact and co-evolve to shape bottled water markets in the UK.

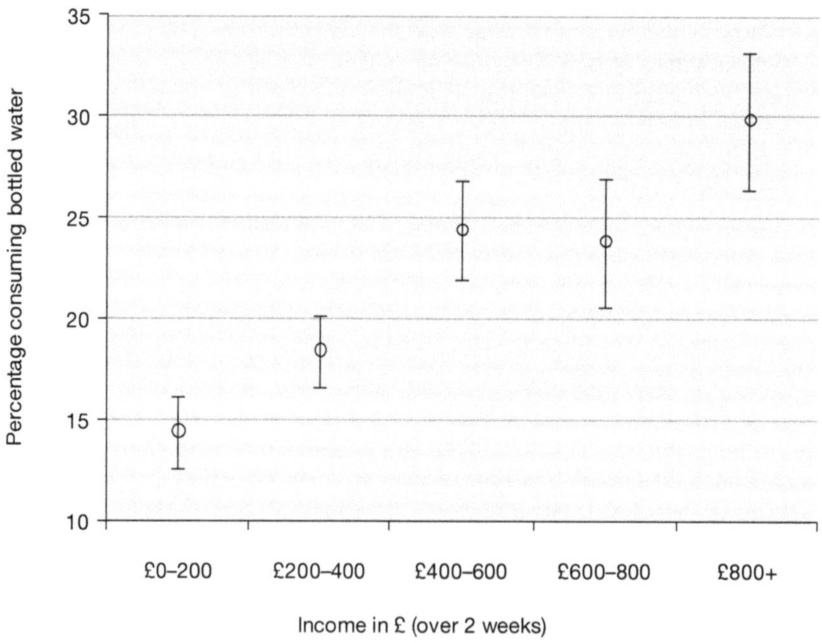

Figure 3.1 UK consumption of bottled water in relation to income (calculated from Living Costs and Food Survey data 2009).

Turning first to consider *processes of consumption*, it is important to note that only approximately 30 per cent of UK households actually consume bottled water (Living Costs and Food Survey, 2009). Furthermore, looking in more detail at socio-demographic and temporal patterns of bottled water consumption reveals some interesting insights.

Figure 3.1 depicts the percentages of households consuming bottled water in relation to income. There is a clear pattern of increasing consumption with increasing income. If the curve had been flatter one could conclude that bottled water is akin to a staple necessity such as bread. If the curve had been steeper one might speculate that bottled water was more akin to a high-end luxury product such as an expensive wine. As it is, the actual relationship is slightly more ambiguous, with clear differences between the lowest and highest income groups but less clear distinctions between middle income groups in terms of their likelihood of consuming bottled water. Nonetheless, bottled water is clearly a relatively marginal discretionary good, complementary to, rather than wholly substituting for, tapwater. The more complex stepped pattern might also be indicative of the complexity of the bottled water market. This is even clearer when later we consider the huge range in price between different bottled water products, the top end clearly identifiable as an extreme luxury good at ten times the price per litre of 'economy' brands (Table 3.4 and following discussion).

The relationship between bottled water consumption and age is also significant (see Figure 3.2). It is clear to see that middle-aged consumers in the UK are on average much more likely to consume bottled water than either their younger or older counterparts. We suspect that this reflects both older consumers' trust in tapwater provision and suspicion of expensive alternatives and their preference for hot drinks and younger consumers' preference for soft drinks such as Coca-Cola.

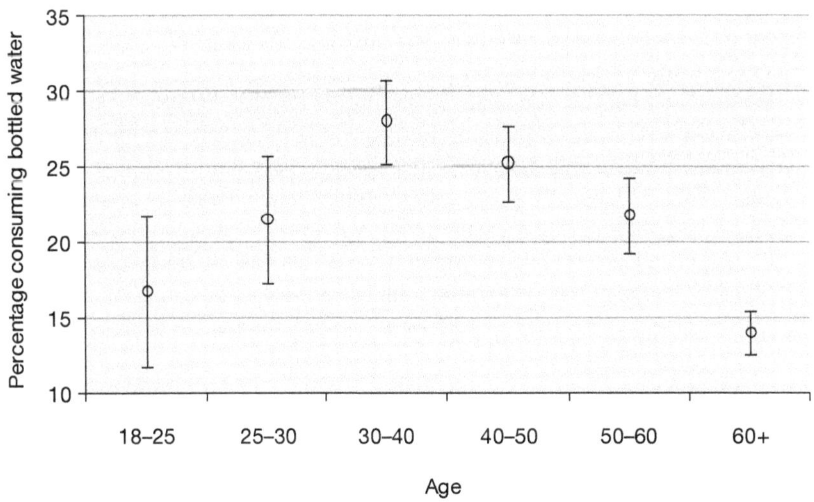

Figure 3.2 UK consumption of bottled water in relation to age (calculated from Living Costs and Food Survey data 2009).

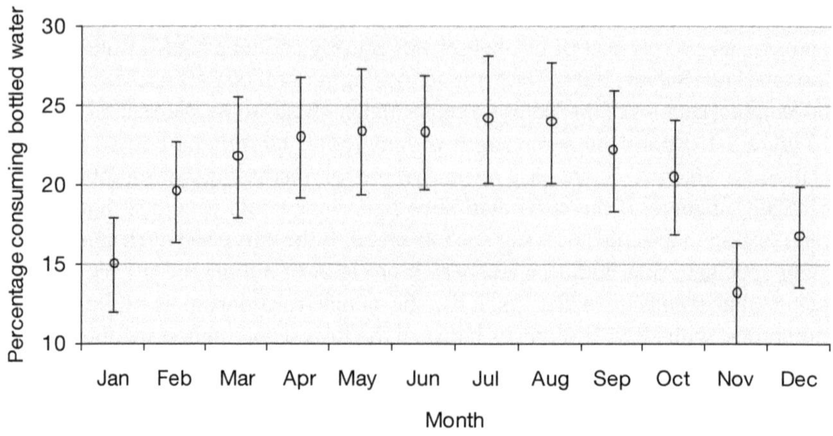

Figure 3.3 Seasonal variations in the consumption of bottled water in the UK (calculated from Living Costs and Food Survey data 2009).

Variations in the consumption of bottled water over the course of the year also demonstrate a significant difference from the routine practices of drinking water from the tap. There is a clear seasonal pattern of consumption with lows in the winter months and highs in the summer months (Figure 3.3). This seasonal effect has also been observed by commentators. For example, as the 2009 Euromonitor report states: *'The bottled water sector continues to be seasonally reliant and changes in the temperature can trigger a noticeable effect. As British summers continue to be mild, bottled water sales tend not to flourish'* (Euromonitor, 2009, 1). This seasonality is one of the defining features of the bottled water and soft drinks industry across Europe and it has significant implications both for the efficiency of production and for labour arrangements. As a key informant observed:

> Seasonality is huge, if you looked at the profile for fizzy drinks – if you go from January it's very low, it picks up for Easter and then there is a big bulge over the summer, then drops down a bit then there's a very sharp peak for Christmas. Other products have got much flatter profiles, that's generally the soft drinks profile, one of the sort of great ambitions was to find a product that would peak in sales in January, February, March time, I don't think anybody ever really achieved that, they looked at milk products, products you can drink hot.
>
> (Retired CEO of a large soft drinks company)

Although there is no doubt some seasonality in drinking water from the tap, these patterns for drinking water from bottles, in various formats, further indicate how bottled water is a different consumption good from tapwater, once placed in the overall context of consumption processes.

If bottled water occupies a distinctive locus in the UK marketplace for soft drinks, it also situates itself in different life and eating styles. People who drink

Table 3.1 Clusters of drink and food commodities (LCF, 2009).

Cluster 1: Fruit and yogurt	Fresh fruit, yogurt, *tea*
Cluster 2: Healthy main meals	Salad, rice, fish, olive oil, *mineral and spring waters*
Cluster 3: Carbohydrates	Biscuits, bread, cake, soft drinks, *fruit juices*
Cluster 4: Meat and potatoes	Bacon, beef, pork, potatoes, *spirits*

bottled water also tend to have distinctive food preferences. Adapting methods developed by Bourdieu (1984) in his analysis of French 'food spaces', it is possible to identify clusters of commodities that are frequently associated with each other to build up a picture of where bottled water fits within these groupings. Principal component analysis was applied to the 2009 Living Costs and Food Survey (LCF) data to identify groups or clusters of commodities.[5] This analysis revealed four fairly distinct clusters of commodities (see Table 3.1).

Our findings clearly indicate that mineral waters form a cluster with healthy products such as salad, rice, fish and olive oil. Interestingly, soft drinks and even fruit juices appear to be more associated with carbohydrate-based snacks such as biscuits and cake. This suggests that UK consumers view mineral water as a healthy product especially compared with soft drinks.[6]

In addition to analysing food survey data, qualitative research with a group of 18 consumers in the UK was conducted in order to gain more in-depth insights into everyday drinking practices. In particular, we used a diary method (see for example Alaszewski, 2006; Buijzen et al., 2008) where people were asked to record details about every drink they consumed on three separate days in a week (Table 3.2).

The first point to note from the UK drinks diaries is that overall there were very few mentions of bottled water compared with other drinks. Only four of the 18 research participants recorded drinking bottled water (yielding a total of 13 separate drinking instances). Moreover, the vast majority of tapwater was consumed in the form of tea, coffee, filtered water and cordials, rather than as plain tapwater. A total of 15 out of 18 diarists drank tea on 150 separate occasions and a total of 14 out of 18 diarists drank coffee on 58 occasions. In contrast only six diarists drank plain tapwater on 19 separate occasions, out of a total of 332 water drinking instances (Table 3.2). Although an obvious point, consumers' daily routines generate the large numbers: cups of tea with breakfast, a mid-morning break, in the afternoon, and so on. Often these routines are socially contextualised, whether in the household or at work. So people drink the same water beverage at the same times, on the same occasions, including both week days and weekends.

5 We used the statistical software package SPSS to conduct the principal component analysis, using the Oblimin rotation method with Kaiser Normalization.

6 The analysis could also be taken further to examine whether these groupings were associated with particular types of consumer but this is beyond the scope of the current research.

Table 3.2 The type and frequency of drinks consumed by UK diary participants.

Drink	*Total number of participants consuming drink (N=18)*	*Total number of times drink was consumed*
Tapwater	6	19
Filtered water	7	31
Bottled water	4	13
Tea	15	150
Coffee	14	58
Squashes and cordials	6	14
Other drinks made with tapwater	8	18
Fruit juices	7	10
Soft drinks	8	19

In IEP terms, a significant aspect of this pattern, albeit from a very small sample of drinks diarists, is that consumers do further work in the home prior to final consumption, transforming the quality of the water to be drunk. Even instances of domestic filtered water exceed instances of drinking untreated tapwater. Finally, the dominant economic form in which tapwater is actually drunk is as a hybrid good: a combination of state-instituted privatised monopoly tapwater, a commodity market modification of it, whether tea, coffee, squashes or other flavourings, and the unpaid work of consumers.

As already noted, only a few of our diarists recorded drinking bottled water, and, with the exception of one of them, infrequently. Even though they were not specifically instructed to do so, they recorded the brand name of the bottled water that they were drinking. This suggests that product brands have significance for them, possibly as an indication of quality. Perhaps more revealing is the fact that bottled water was consumed in a variety of different contexts and associated with a range of different everyday practices, such as working, exercising, travelling etc. Diarists also identified a range of different motivations for drinking bottled water, from socialising and relaxing to cleansing the palate. Only one diarist who accounted for nine of the 13 drinking instances, drank bottled water regularly, routinely accompanying meals, sometimes with wine. These associations are far from trivial and any attempt to understand the demand for bottled water, or indeed to manage that demand, would need to account for the different ways in which bottled water is being consumed. The diaries incidentally reveal that consumers were concerned about the environmental implications of bottled water consumption, especially in relation to plastic bottles. In contrast, fewer concerns

Table 3.3 Terms used by UK research participants for tastes of different drinks.

	High usage	*Medium usage*	*Low usage*
Soft drinks	Refreshing Delicious	Cold Sweet Chemical	Thirst quenching
Tapwater	Refreshing		Chemical Metallic Chlorine No taste/Bland Thirst quenching
Bottled water	Bland/Non-descript	Refreshing Delicious	General positive Cold Cleansing palate Duty/Habit
Filtered water	Refreshing Delicious	Bland/Non-descript	General positive

were raised about the consumption of soft drinks or hot drinks, which tells us something about the relative profile of these products in environmental terms. Yet, as we have seen, soft drinks are more environmentally damaging and grab a much larger market share in the UK.

Research participants were also asked to record the taste of the drinks that they consumed (see Table 3.3).[7] Tapwater was frequently described as 'refreshing', however terms such as 'chemical', 'metallic' and 'chlorine' were also used, albeit less frequently. One diarist even recorded throwing away some of their tapwater, as the taste was so unpleasant. This gives us a fleeting insight into perhaps one of the most important factors relating to the growth of the bottled water market in the UK, namely the perceived quality and safety of UK tapwater. Despite the assertions of authors such as Gleick (2010), it is clear that for many UK consumers bottled water is perceived as having some genuine 'points of difference' from tapwater, primarily in relation to quality and taste. A second point relates to the descriptions of the taste of bottled water. Whilst bottled water did not receive any negative comments about its taste, it tended to be described in fairly mundane terms such as 'bland', 'cold' or 'refreshing'. One might speculate that, due in part to the immaturity of the bottled water market in the UK, consumers have not yet developed the competencies or vocabularies to describe the tastes of these products, or to differentiate between different mineral waters.

7 For a discussion of the ways in which taste and other aesthetic and embodied dimensions of food/drink consumption intertwine with ethical considerations, see Evans and Miele, 2012; Miele and Evans, 2010.

Turning to *processes of production* of bottled water in the UK, three defining characteristics stand out in the organisation of production. Firstly, 'private labels', the supermarket brands, make up the largest single share (38.8 per cent) of the UK bottled water market in 2008 (Euromonitor, 2009). This reflects the economic power relations between retailers and manufacturers in the UK, and the dominance of supermarkets. This is interesting in itself as previous academic literatures have tended to focus on the role of branded bottled water products and their associated marketing strategies (see Wilk, 2006; Brei and Böhm, 2011), rather than supermarket brands, which have very little dedicated marketing or advertising strategies. The case for the consumer dupe is seriously undermined, in the absence of the blandishments of advertising and aggressive marketing.

Producers tend to specialise in either private label production (e.g. companies such as Princes) or branded production, with fewer companies (such as Highland Spring) offering both branded and private label products. This is in part due to the potential conflicts that can emerge when a single company produces both branded and private label products at the same time, particularly if no clear quality differences between them can easily be established. This is perhaps even more the case with bottled waters than with pizzas, soups or pies (see Harvey et al., 2002).

> It's quite difficult to do a branded product and a retail own label at the same time, it's fine if you're just using a bit of spare capacity but if you're doing it on a large scale then you've got to have the capability to provide both, so if you're packing your product and you're packing for Tesco then all of a sudden you get a really hot summer, sales go up through the roof, demand soars, who do you pack for, do you pack your own product, or do you pack Tesco's product?
>
> (Representative from the British Soft Drinks Association)

The fact that a large percentage of bottled water production in the UK is private label or supermarket 'own brand' has several implications for the overall character of the market. Indeed, it places a large section of the bottled water market entirely under supermarket regimes of product qualification and pricing and it highlights the overall power of the multiples in configuring bottled water markets.

Secondly, the branded production of bottled waters is concentrated in a small number of large companies. In 2008 the top 10 UK bottled water companies, including Danone, Highland Spring, Nestlé, Silver Spring and Barr, had a 50 per cent share of the total market for bottled water. The main reason for this market concentration seems to be the advantages offered by economies of scale especially in the context of a highly integrated and concentrated national retail market.

> A lot of these things come down to scale and the size of the company, you'll find that the biggest companies … [have an advantage] in as much as they could dedicate two lines just to … [a single product] they would never have to have a product change over on those lines – they could just keep those lines belting away if necessary 24 hours a day producing one product.
>
> (Retired CEO of a large soft drinks company)

The size and concentration of production has knock-on implications for the types of water supplied and, in particular, the speed of product innovation, as one key informant remarked:

> … they [bigger companies] are less flexible than we are, so our NPD program and our speed to market, if we spot an idea, is in their terms like a nano-second. They can take over a year to decide something and I can get something to market in eight weeks.
>
> (CEO of a small soft drinks company)

Thirdly, it is important to note that many of the bottled water producers in the UK are owned by soft drinks companies. This was emphasised by one of our interviewees from the British Soft Drinks Association:

> Whereas on the continent you tend to find companies that do soft drinks and companies that do bottled waters. Here I think all of the soft drinks, certainly all of the major soft drinks companies, do bottled waters as well.

This market characteristic has significant broader implications for the debate regarding health and environmental issues around drink consumption in the UK. Some bottled water manufacturers appear unwilling to contrast their products with soft drinks, even though they might compare favourably in both environmental and health terms.

Turning to *processes of distribution and retail* in the UK bottled water market, the vast majority of bottled water is sold through large multiple-store retailers. As one key informant observed, producers have to focus their attention on a few key market players:

> 85% of the grocery trade goes through five or six major multiples, if you're launching a new product you've only got to sell it to about five people … the buyer at Asda, the buyer at Tesco, the buyer at Sainsburys …
>
> (Representative of the British Soft Drinks Association)

In some ways this has many advantages for the larger bottled water producers, especially in terms of guaranteeing a certain level and consistency of demand, at least in the short-term. However it also places a lot of power in the hands of a few key supermarkets and their buyers. The difficult and ambiguous nature of the relationship between drinks producers and supermarkets was summed up by another key informant:

> … they are our route to market. Without them we would not have the business we have, we then have to try and build a relationship and a way of trading that works for us and for them … We have to play to our strengths, we have to be knowledgeable of and wary of the power of them.
>
> (CEO of a small drinks company)

Due to the power and market share of the supermarket sector in the UK, supermarkets are able to exert enormous pressure on bottled water producers. In particular, supermarkets significantly influence product prices, product margins, and control producer promotional offers. They also request additional payments for in-store promotions and shelving/displays (e.g. gondola end displays). To a certain extent, high profile branded producers can counteract supermarket power, exercising some control over issues such as product price. The situation was more bleak for dedicated private label producers.

> If you are making just own label products, then you have no control over the price whatsoever and in fact most of the major retailers now they put their production up for auction. I'm not sure if they still do it, they certainly did – internet auctions – and manufacturers and bottlers will bid, they say we will supply you at so many pounds.
>
> (Representative of the British Soft Drinks Association)

A second interesting feature of the distribution and retail of bottled waters relates to the range and variety of bottled waters available in UK supermarkets.

There are significant differences between the levels of variety on offer at different UK supermarkets. Looking at Table 3.4, we see that Waitrose supplied no less than 67 different types of still water through its internet shopping service in 2012 and 35 different types of sparkling water, and in our market audits, in-store displays often had 30–40 varieties. But if you are a regular shopper at Asda, your experience of the bottled water market would be quite different from that of a regular shopper at Tesco or Waitrose. Some of this variety can be explained by retailers offering the same product in a range of different bottle sizes and package configurations. However, these volume differences are not trivial, as smaller bottles are often sold for a much higher price per volume than their larger equivalents, as consumers seemed willing to pay significant premiums for portability and convenience.

Other sources of variety related to water type (still, carbonated, flavoured, functional), brand type (branded, own label), origin, legal classification (natural mineral water, spring water, bottled water), bottle design (e.g. shape, sports cap), bottle material (glass, plastic) and format (e.g. individual, multi-pack). Each of these product features are tightly articulated with the price of bottled waters in different ways, leading to a very large price range between the cheapest and most expensive products. For example, at the time we conducted the market audit, the most expensive still bottled water available at Waitrose was a six pack of 250ml Aqua Panna, which retailed at 188p per litre, whereas the cheapest was a Waitrose own brand natural mineral water, which retailed at 18p per litre. Quality differentiation and price differentiation go hand-in-hand, and there is no free-standing 'economy of qualities' (Callon et al., 2002; Heuts and Mol, 2013). There is clearly a quality–price matrix, ranging from the high-end luxury good to the relatively modest discretionary quality alternative to tapwater.

Table 3.4 Varieties of bottled water for sale in UK supermarkets 2012.[8]

	Asda	*Tesco*	*Waitrose*
Still	32	43	67
Sparkling	13	15	35
Flavoured	34	31	23
Functional	3	11	7

What these results make very clear is that it is not possible or desirable to treat bottled water as a single unified product in the UK market. Bottled water is an umbrella term for a variety of commodities. It can be viewed as a dedicated drinking water qualitatively superior to the consumer's local chlorinated tapwater, as an alternative to soft drinks, or even as an ostentatious luxury complementary to drinking an expensive wine. It is also clear that different types of bottled water products are engineered, designed and marketed to fit with particular sets of drinking practices, for example glass water bottles fit with luxury consumption, sports caps offer convenience and portability and larger containers provide the necessary bulk for bottled water to be used as an everyday complement to tapwater.

Although hidden from sight and little discussed, another key feature of the UK bottled water economy concerns ownership and appropriation of water resources – the source that is the precondition for the sip. In Chapter 1, a distinction was drawn between ownership, or rights of control, over water resources and water provisioning. It was argued that the development of the institution of water resources as a UK national public good occurred historically quite late, in the 1960s and 1970s. In the case of bottled water, as the most evidently commodified form of water even compared with privatised tapwater, the key issue is one of the rights of commercial companies to abstract water from the national public water resource.

Bottled water is particularly Janus-faced in relation to commodification. On the consumer-side it is very clearly a commodity to be bought and sold; enclosed by a bottle; categorised now by an EU regulatory regime; and priced by companies often in negotiation with retailers, purchased and transported by consumers (or ordered by internet and delivered to the home). In contrast, prior to abstraction, as the 'common pool' water resource occurring naturally in the ground, it is definitely not a commodity. As discussed in Chapter 2, the process of forming public property rights over surface and groundwater is a complex, often conflictual, historical one, varying in different societal contexts. In the words of one key informant:

8 This data is from a market audit we conducted in 2012 looking at the availability of different bottled waters. The audit was conducted using information from the internet shopping sites of these retailers.

> The state licenses its use but the principle of licensing isn't about … it's got nothing to do with ownership it's all to do with controlling the impact of use rather than you know … well I'm going to sell you the water.
>
> (Consultant to the bottled water industry)

In the UK, the Environment Agency now controls abstraction and requires a license for any abstraction over 20,000 litres (or 20 cubic metres) per day (Environment Agency, 2008). The fees for these licenses are currently minimal, indicating that the relation between the state and bottled water manufacturers is not one of a market exchange, with a market price assigned to the yet-to-be abstracted groundwater. Rather the state is exercising control over rights to abstract and quantities permitted to be abstracted. As we will see in the case of Taiwan, different states can set very different fees to private companies for the license to exploit national water resources As such, the license fee is a kind of litmus test of a state–market relationship in a given society. In the UK

> It's an administration charge and it's peanuts, it's done on the basis of the volume you use and the use but it's not very much … let's say you're producing 500 cubic metres a day the costs of operating, even powering the borehole pump is going to exceed the license fee … It will be hundreds of pounds a year … They're very low charges, it varies region to region but it's not a major part of a company's cost.
>
> (Consultant to the bottled water industry)

There are occasional examples of landowners owning an abstraction license who exploit a particular groundwater source and sell the water to bottled water companies. The Buxton abstraction license, for example, is held by the Local Authority and the water is sold to Nestlé, an interesting case of a different state–market exchange. But this practice is more common on the continent than in the UK. Even when this does occur, the amount charged for the water is often little more than what one would pay for mains water supply.

A further key aspect of property rights over water abstraction for bottled water companies is that many licenses to abstract have reached their capacity limit. At the same time, many licenses are historical and their owners do not use their full quota of abstraction. Consequently, the Environment Agency has been keen to promote the trading of abstraction licenses to free up capacity. If a bottled water producer needs to develop a new source in the UK, it is increasingly likely that they will have to purchase a license from a current holder rather than be granted a new license from the Environment Agency. The price of the license then becomes a market price for exchange between two market players. The Environment Agency, in permitting the trading of licenses, has created a market for license-transfers, in contrast to the non-market granting of original license assignments. Public and private forms of ownership and exchange thus intertwine in distinctive ways for the exploitation of natural water resources, as a precondition for the national bottled water industry in the UK.

The Dasani episode. To end our discussion of bottled water market configuration in the UK, the failed attempt to introduce a processed bottled water, Dasani, reinforces the preceding analysis (Wilk, 2006; Gleick, 2010). In early 2004 Coca-Cola launched a bottled water product, Dasani, in the UK market. Despite the fact that it had previously enjoyed good sales in other markets, particularly in the US, the product failed to make any impact in the UK market and within a few months of its release it was withdrawn from sale. The discovery of higher than permitted quantities of the suspected carcinogen bromate in the water was the immediate pretext for its withdrawal. One cannot begin to understand the failure of Dasani in the UK without including in the analysis a wider range of different factors and especially their interaction.

Firstly, contrasting cultural norms and expectations between UK and US consumers and producers (and more specifically, production methods) underlie the failure of Dasani in the UK. Unlike the vast majority of bottled waters already on sale in the UK, Dasani was processed water, produced by reprocessing ordinary municipal tapwater supplies using reverse osmosis technology. This method of production did not sit well with UK consumers and sections of the UK media who had strong ideological preferences for bottled water directly sourced from 'natural' groundwater. It also raised questions about large corporations making profits from public water supplies. As one of our key informants stated:

> I just remember it from a consumer point of view and it just looked like they were trying to sell something that comes from a tap, trying to pass it off as a naturally sourced water. Given the price that they were charging for it and the packaging, it did suggest that it was not municipal source but natural source.
>
> (Representative of the UK Natural Hydration Council)

Secondly, even more crucially, different bottled water legislation in the US and Europe led directly to the contamination of Dasani with trace elements of bromate. Legally, Dasani in the UK could not be the same drink as Dasani in the US because there is a legal requirement for UK drinking water to have a minimum level of calcium. Coca-Cola had been merrily producing Dasani bottled water in America without any bromate issues. Problems only arose in the UK because the reverse osmosis processes removed naturally occurring calcium from tapwater. Calcium then had to be artificially added back into the water. This in itself would not have been a problem, except for the fact that Coca-Cola mistakenly added calcium in the form of calcium chloride, resulting in the bromate contamination as explained by a UK market expert:

> Dasani is massive abroad but when they tried to do it in this country in the UK you have a regulation whereby any processed water has to contain a minimum of 60mg/l of calcium … in order to add the calcium they added it as calcium chloride and calcium chloride can contain very minute traces of calcium bromide … and that gets oxidised to bromate and the limit for bromate in processed water is extremely low.
>
> (Representative of the British Soft Drinks Association)

If it had not been for the specific UK legal requirement for calcium, coupled with the company's lack of familiarity with this process, then the bromate might have never found its way into the water in the first place. Legislation and technologies of production intertwined in a curiously entangled fashion to produce a market disaster. The Dasani episode thus provides a useful illustration both of the distinctiveness of the UK market compared with the US and of the complex interconnections between processes of qualification of water, markets and norms of consumption. In particular, it highlights the quality contrast instituted within the UK market between industrially processed and/or reprocessed tapwater and bottled spring waters abstracted from underground sources. As we shall see in the case of Taiwan, a quite different dynamic of quality distinction can be created between a standard industrially processed tapwater and high-tech reprocessed and synthesised water. But societal variation over qualities of water matters even within the narrower frame of Europe, as we shall now see in the two contrasting cases of Italy and Germany.

Nationalist and localist: the Italian bottled water economy

Italian consumers drank 10.2 billion litres of bottled water in 2009, which equates to approximately 0.4 litres per person per day (Euromonitor, 2009). This makes the Italian per capita bottled water market one of the biggest in Europe and *four times* bigger than the UK market. The majority of bottled water consumed was still water (5,762 million litres in 2009, Euromonitor, 2009) but a significant amount of sparkling water was also purchased (2,556 million litres in 2009, Euromonitor, 2009). Over the coming sections we draw on a range of empirical research, including key informant interviews with producers, retailers and regulators and diary research with Italian consumers to present a more in-depth account of the nature of bottled water markets in Italy. As in the UK case study, we will focus on the key economic processes of production, distribution and consumption that interact to configure bottled water markets in Italy.

In terms of bottled water production, the Italian bottled water market is even more concentrated than the UK, the top ten companies having a 78 per cent market share. Unlike the UK branded bottled water companies, Italy is dominated by nationally named companies, and, with the exception of San Pellegrino, the companies are also nationally owned. The top four bottled water producers in Italy in 2009 were San Pellegrino (Nestlé) with a 25 per cent market share, San Benedetto Acqua Minerale with 18 per cent, Ferrarelle (Vitasnella, Boario, Santagata, Natía) and CoGeDí International (Uliveto, Rocchetta), each with 9 per cent. Many of these national brands produce local and regional brands. In addition to these very large bottled water companies, a large network of smaller producers supply more local food retailer networks.

In direct contrast to the UK market, only 6.5 per cent of production in Italy in 2009 was dedicated to the supermarket private labels (Euromonitor, 2009). Nevertheless, these private label producers (many of which tended to focus exclusively on private label production) were considered to occupy an important and potentially growing market niche. As one key informant stated:

> We have a number of springs that are just sourcing for private label … The 'saint waters' we call them because they are from Saint Antonio, Saint Carlo … They have a lot of Saints and they supply a lot for the Co-op and Conad etc.
>
> (Italian bottled water producer)

In addition to the very large companies which dominate the bottled water market in Italy, there are also a large number of smaller producers who supply at a more regional level and who are connected with more local food supply networks. Furthermore, even the few big multiple grocery outlets, such as Co-op Italia, are keen to source their food and drink from regional suppliers and, as we shall see, one of their recent ethical campaigns focused attention on the distance different bottled waters had to travel before arriving in their stores. There is thus a distinctive nationalist and localist flavour to Italian bottled water. This characteristic was indeed borne out by our Italian drinks diaries, where 100 per cent of consumed bottled water was Italian, and mostly from the diarists' own region. This contrasts with the UK, where two of the three recorded brands drunk by our diarists were foreign brands.

Turning to processes of *distribution and retail*, in 2009 71 per cent of all bottled water sales were in supermarkets and discounters, 10 per cent were classed as 'traditional and delivery' and 19 per cent were sold through catering outlets and vending machines. But the power relations between producers and retailers are very different in Italy, exemplified by the dominance of manufacturer brands and the tiny market segment for private label. Bottled waters were mainly sold through small and regional supermarkets. Even the largest retailer in Italy (Co-op Italia) is regionally managed and has far fewer outlets in the South of Italy. As such, supermarkets in Italy do not have the same bargaining powers as their UK counterparts. The power tensions between producers and retailers in Italy were revealed in interviews on both sides of the divide over whether producers or retailers were able to wield the most power and influence in shaping the market for bottled water. The producers that we interviewed told a similar story to their UK counterparts about supermarkets demanding money for special offers and promotions:

> To go in a chain of supermarkets you have to pay and then if you want to make promotions, normally they ask for money you know the paper that they send home the advertisement that they make in a magazine or newspaper they ask all the people who are on the advertisement 500 Euro, 10,000 Euro it depends.
>
> (Italian bottled water producer)

In contrast, the retailers that we interviewed told of the enormous power of bottled water companies, who 'commanded advertising budgets second only to the car manufacturing industry' (President of a regional supermarket chain) and who provided one of the key commodities attracting consumers to their stores.

This dominance of major national and local brands also reflects the more mature nature of the bottled water market in Italy and the fact that many Italian consumers are dedicated to particular branded or local varieties of bottled water. The market audit of Italian retail outlets selling bottled water revealed similar amounts of diversity and variety of bottled waters as in the UK. However, unlike the UK, bottled waters often occupied far more prominent positions in stores, indicating their status amongst other products. Bottled water is one of the key commodities sold by supermarkets:

> Like Nutella and Coke, water products have to be there as part of a range. If they are not in the store then people will literally start to shop elsewhere.
>
> (President of a regional supermarket chain)

The bottled water market in Italy also appeared to be very price sensitive and there was considerable competition between brands in relation to price and special offers. Finally, the most common size of bottled water for sale in Italy consisted of larger containers of 1.5 to 2 litres or more, which were frequently sold in multi-packs (75 per cent of the market, according to one retailer we interviewed). This bulk provisioning of bottled water hints towards its usage as an everyday replacement for tapwater for drinking. In many urban areas, moreover, there is an extensive home delivery service for bottled water.

And indeed, Italian consumers do seem to be widely using bottled water as a direct substitute for tapwater, a distinctive feature of *consumption* within the overall Italian configuration. When we compare our 15 drinks diarists (all inhabitants of Piombino in the province of Livorno), the contrast with the UK is remarkable, as is evident from Table 3.5.

Table 3.5 The type and frequency of drinks consumed by Italian diary participants.

Drink	*Total number of participants consuming drink (N=15)*	*Total number of times drink was consumed*
Tapwater	1	1
Filtered water	0	0
Bottled water	15	193
Tea	8	28
Coffee	14	92
Squashes and cordials	0	0
Other drinks made with tapwater	0	0
Fruit juices	6	9
Soft drinks	5	7

When Italians drink plain water, they primarily drink bottled water (193 occurrences for the 15 diarists over a three day period). Coffee (92 occurrences) comes second, but given the way Italians take their coffee (espressos), this probably represents much less in terms of relative quantity. Tea is drunk frequently (28 occurrences), but, as one would expect from cultural stereotypes, not remotely on the scale observed in the UK. Finally, the consumption of tapwater, in plain form direct from the tap, was very rare (one occurrence). In several Italian regions there is a lack of consumer trust in the safety of tapwater due to past pollution incidents and regional problems with naturally occurring arsenic and uranium. Moreover, there might well be a particular bias in our sample of diarists. Interviews with the local municipal authorities revealed that historically there had been serious natural hydro-geological contamination of water sources in Piombino, although the local tapwater supply is now affirmed to be safe to drink.

> Historically arsenic and boron are big problems in Piombino and the surrounding area … 33 municipalities share the same catchment area and they have grouped together to form a consortium.
>
> (Piombino council member responsible for sustainability)

Recently, an EU grant had enabled more investment in water treatment facilities in order to reduce the levels of naturally occurring arsenic to within permitted EU standards.

However, in saying that water is drunk from the bottle rather than the tap, it is also significant that bottled water is overwhelmingly spring or mineral water, not processed water. The aesthetic and mineral qualities of water are highly valued by consumers, qualities that tapwater could not deliver. In that sense, even in conditions of lack of trust for tapwater, bottled water is not a substitute but an alternative beverage. The preference for Italian, regional and even local waters with distinctive tastes and mineralities is an important feature of the Italian bottled water market. The diaries also showed that individual consumers often drank several different bottled water brands over the course of a day. This was partly because they drank what was available at different locations (the homes of friends and family, work places etc.) and partly because they often kept more than one brand in their own homes.

Furthermore, in the same way that there is evidence to suggest that drinks such as tea have become embedded in broader UK consumption practices and routines (especially rituals around eating and socialising) there is also evidence to suggest that the consumption of bottled water has become entrenched in certain aspects of the Italian way of life. Many diaries showed regular patterns of repeated bottled water consumption over the three day period, for example one family always had two bottles of mineral water (one sparkling and one still) at almost every meal time. This indicates that, for this family at least, drinking bottled water had become indelibly associated with broader meal time rituals.

In addition to regularly consuming bottled water at meal times, the Italian diarists also consumed bottled water in conjunction with a range of other routine activities, including whilst working, socialising and playing sport. One diarist noted that consuming bottled water had become part of her health routine. Yet, in spite of the large quantities being consumed, our evidence suggests that, as in the UK, some Italian consumers were concerned about the potential environmental implications of bottled water consumption. Many diarists commented on the environmental implications of plastic water bottles and the need to recycle them where possible. The Italian diarists also drank fruit juices or soft drinks infrequently, primarily due to such drinks being perceived as having an unhealthily high sugar content. The market for soft drinks, fruit juices and tea in Italy is small in comparison to the bottled water market.

Summing up, this analysis of the instituted economy for bottled water in Italy differs significantly from the UK across each of its dimensions, how it is produced, distributed, sold and consumed. The societal contrast is between the Italian and UK configuration as a relational whole. As we will discuss in concluding the chapter, there is a quite different relation between the economies of bottled and tapwater in the different societal contexts. Despite the near-universal provision of public tapwater in Italy, tapwater is not primarily used for drinking plain water.

Interestingly, however, there are some emergent signs of this dominant cultural and economic dichotomy between tapwater and bottled water being destabilised. Recently municipal authorities have made significant efforts to highlight the improved safety of tapwater and to promote the use of publically supplied water through the provision of public water fountains, some of which even dispense chilled and sparkling waters. Ethical supermarkets, such as Co-op Italia, have become one of the most important voices questioning the environmental credentials of bottled waters. Recently, Co-op Italia launched a campaign called '*Acqua di casa mia*' or 'water from my home'. This campaign tied into key food sensibilities in Italy about the importance of local cuisine. It also played off the meaning of 'water from my home', which was meant to simultaneously imply that consumers should be drinking tapwater or at least consuming local bottled waters, rather than bottled waters that had travelled from further afield either in Italy or abroad. This campaign seemed to strike a chord with consumers, who were also able to look up exactly how far their branded water had travelled using in-store poster displays.

However, as one might imagine, the campaign did not sit very well with bottled water producers, one of whom expressed his concerns about the promotion of tapwater, which he viewed as potentially unsafe. For him, the bottled water market was underpinned by distrust in tapwater – which is what he wanted to promote.

> Even the Co-op started last year with an advertisement, don't drink mineral water – the tap water with the carafe is very good. So we are really scared about this because a carbon filter can just take out chlorine from the taste. But you can't change the chemistry of the water. Our chemistry is really a serious thing requiring many analyses during the year.
>
> (Italian bottled water producer)

The campaign also represented a big risk for Co-op Italia, as it lost a purported 20–30 million euros in direct payments from bottled water companies as a consequence, quite apart from sales of its own private label bottled water. The future is open, and instituted economies can always be de-instituted, and then reconfigured.

Sparkling with distinction: German bottled water economy

Consumers in Germany drank 10,518 litres of bottled water in 2008 (Euromonitor, 2009), which amounts to approximately 0.35 litres per person per day. This makes the German per capita market similar in size to the Italian market and three and a half times the size of the UK market. Perhaps one of the most striking features of the German bottled water market relates to the amount of highly mineralised sparkling water consumed. Unlike in the UK and Italy, German consumers drank far more carbonated than still bottled waters. Indeed, in 2008, Germans drank 7,624 million litres of carbonated water, compared with 1,928 million litres of still water (Euromonitor, 2009). One German consumer we interviewed even stated that they did not feel fully refreshed unless they had drunk sparkling water. There are many possible explanations for this preference for sparkling water, which is a key feature of the German market. First, on account of hydro-geological characteristics there are several naturally carbonated springs in Germany, which produce good quality, highly mineralised waters. Second, because carbon dioxide has a natural sterilising effect it is likely that these mineral waters would have been highly valued, especially historically when water treatments were less sophisticated than today. There has been a long history of spas and the consumption of different mineralised waters for specific therapeutic purposes in Germany (Back et al., 1995; George, 1994). Indeed, the concept of 'mineral water' was first chemically defined at the International Balneological Congress in Nauheim, as containing a minimum of 1000mg of dissolved minerals per litre, exceptionally high even by European standards (Diduch et al., 2011). Third, bottled water companies such as Gerolsteiner go to great lengths to promote the health and quality benefits of consuming highly mineralised sparkling waters. Overall, Germany manifests a quite distinctive combination of high quality tapwater and sparkling mineral water. One thing is certain: they are not the same drink, and they are not rivalrous products.

In terms of *production*, the German bottled water market is much less concentrated than in the UK or Italy. The top ten companies had only a 32 per cent market share in 2008 (Euromonitor, 2009) and the top three bottled water producers in Germany were Nestlé (11 per cent), Gerolsteiner Brunnen (9 per cent), Danone (6 per cent), with the remainder each having less than 4 per cent, supplying often very small and local markets. A defining feature of the German bottled water market relates to the division between small-scale local producers, often selling to specialist drink shops and large-scale producers specialising in private label production for German discount supermarkets, such as Lidl and Aldi. Private label production constituted 22.3 per cent of total bottled water production in Germany in 2008 (Euromonitor, 2009), far lower than the UK at 38.8 per cent. German food and drink retailing is characterised by an almost schizophrenic divide between price driven bulk-quantity discounters, and quality-distinguished traditional small outlets (Kjaernes et al., 2007). Key informant interviews revealed competition and distrust between the traditional producers and the private label producers, who were often disparagingly referred to as 'no-names'.

> There are some really big companies, usually delivering or producing and delivering for hard discounters like Aldi, Lidl, Netto, Penny. That is more than half of the market for the time being and the rest is the normal market brands ... We describe the hard discount as 'no-names' or retailer brands and the rest is brands.
>
> (German bottled water trade organisation representatives)

Many of the smaller, traditional bottled water producers belong to the trade company GDB (Genossenschaft Deutscher Brunnen) and the affiliated trade organisation VDM (Verband Deutscher Mineralbrunnen).

> We have about 200 members in VDM, so still a lot, mostly small and very small companies, family owned companies, 4th, 5th, 6th generation ... which is a speciality for us as an organisation.
>
> (German bottled water trade organisation representatives)

One of the distinctive features of this arrangement is that GDB organises a reusable bottle and crate 'pool' system for producers. All producers who are members of GDB share the same distinctively shaped dimpled bottles, standardised except for removable bottle labels. This arrangement has a series of important implications both for the sustainability of the product, the logistics of storing, cleaning, inspecting and redistributing reusable bottles, and the qualification and marketing of the product in store.

> We are a co-operative society ... we have member companies that own shares of the GDB ... and we are working for these member companies as a service organisation today ... we are supplying all the goods they need to run their business, from water pumps to chemicals, to advertising material ... but the main turnover is by packaging, that means closures, bottles and crates.
>
> (Representative of GDB)

This organisation of producers in turn has significant implications for the different channels of *distribution and retail*. Discounters (such as Aldi and Lidl) and supermarkets sell the majority of bottled water in Germany at 50 per cent and 19 per cent market share respectively. However, specialist beverage stores, which often stock more local bottled water brands, account for 10 per cent of bottled water sales (VDM, 2011). The discount supermarkets sell their own label products in PET non-recyclable bottles, while in contrast specialist drinks stores often sell local water brands in re-usable glass and PET bottles. An interesting feature of the German bottled water market relates to the fact that consumers pay deposits of varying levels on the majority of bottles, even non-reusable PET. This market institution directly affects the pricing of German bottled waters, as consumers are only effectively paying for the water itself, rather than for both the water and the container, as in UK and Italian markets. This feature at least partly explains the significantly lower prices for the high quality mineral and sparkling bottled waters in Germany, especially when compared to the UK.

Unlike the UK and Italy case studies, the collection of drinks diaries was unfortunately beyond the scope of this research. Relying on market data, however, it is clear that for German consumers, sparkling mineral water represents a basic product, in contrast to the UK where sparkling mineralised water is priced and branded at the high end of the market. Even the discount supermarkets predominantly sell sparkling mineral water. As already noted, there is a long history of consuming highly mineralised, usually sparkling waters in Germany. Moreover, German distinctiveness was highlighted in 1980 by a conflict over the German qualification of 'mineral water', and other European countries such as France and Italy, when the European Union imposed a new Europe-wide qualification defining mineral content.

> Prior to 1980 in Germany and Eastern Europe to be classified as a mineral water, a water must have a high mineral content of at least 1000mg dissolved solids … and to have this content it has to be sparkling … in 1980 the EU directive on natural mineral waters tried to put together two different animals under one legislation and this created a lot of problems … prior to 1980 most French bottled waters could not be called mineral water in Germany.
>
> (German bottled water trade organisation representatives)

Significantly, the process of producing this water quality requires carbonation in order to dissolve these high mineral concentrations, according to key industry informants. German consumption is indeed sparkling with distinction. In Germany, mineral waters have a very large (46 per cent) market share of the whole non-alcoholic cold drink sector, compared to 6 per cent for all bottled water in the UK. Moreover, many soft drinks are made using mineral waters (this category has a 15 per cent share of the non-alcoholic drinks market). The share of 'other soft drinks' (26 per cent) is markedly lower than in the UK. The market for less mineralised spring/table waters is very small in Germany (2 per cent) (Taylor Nelson Sofres, 2010; Verband Deutscher Mineralbrunnen, 2011). Thus, German consumers have a quite distinctive overall pattern of drinking bottled beverages, compared to both the UK and Italy.

Compared with other European countries, German bottled water is distinctive both in its own sparkling quality terms, and in relation to its tapwater quality. German consumers appear to have a high trust in the safety of tapwater. It is perceived to be high quality and good tasting in many areas. This is partly because German authorities have adopted an ecosystem/watershed wide water management approach aimed at preventing water pollution and ultimately reducing the amount of chemical treatment required to make surface and groundwater drinkable. As a representative from the German Federal Environment Agency informed us:

> We've got about 20% of the German supply by surface water sources and they tend to get disinfected [with chlorine]. But the groundwater often is not and I think about 50% of Germans are supplied with drinking water that is not regularly disinfected, because soil filters out the bugs so nicely.
>
> (UBA – Federal Environment Agency, water quality department)

Nevertheless, despite the perceived safety, quality and taste of tapwater, many consumers have a strong preference for bottled carbonated waters with a high mineral content, for reasons of both health and taste. Bottled water is drunk in positive rather than negative preference in relation to tapwater.

Conclusions: contrasting configurations

Comparing our three cases, perhaps the most striking aspect is how different are the *overall* configurations of bottled water economies in the three countries. The way that processes of production, distribution, exchange and consumption all hang together relationally in these different societal contexts is at least as remarkable as the contrasts between each of the processes viewed separately. For example, it is almost unimaginable in the UK context to find the kind of cooperation between brand manufacturers in sharing the same bottles for recycling that occurs in Germany. Yet that difference in organisation of producers only makes sense in the context of the German retail organisation, and also the overwhelmingly homogeneous pattern of consumption of sparkling mineral water. Finally, it connects to exchange: in Germany consumers buy just water and in the UK, bottles with water inside.

Moreover, each of these cases exhibit a different complementarity with tapwater: safe but more highly and variably chlorinated surface water in the UK; often less chlorinated and groundwater in Germany; and less safe through natural contamination in Italy, at least in consumer and expert perceptions. Although we have not been able to explore this issue in depth, these different water economies, both bottled and tap, also seem to reflect very different environmental conditions for resourcing water for drinking in our three country cases. Different bottled water economies inhabit different natural environments.

Bottled water for sale in the UK is fundamentally not the same as the bottled water for sale in Italy or in Germany – they are different products in differently organised market spaces. Crucially, they are not just different products in the sense that German bottled water is often highly mineralised and sparkling and that UK water is frequently still. They are different products in the sense that consumers consume them in different ways within their different configurations. Thus for example, the highly routinised and habitual purchase of locally sourced 1.5 litre bottles of still water in Italy as a substitute for potentially unsafe tapwater is very different from the occasional consumption of imported bottled waters in the UK as an alternative to soft drinks, which in turn is very different from the situation in Germany. Not only are the drinks different in their consumption use, but also in the ways in which they are qualified and priced and in the retail contexts in which they are purchased.

This configurational approach demonstrates clearly the inadequacy of viewing consumer demand for bottled water in terms of individualised preferences and behaviours. Instead, understanding the demand for bottled water, particularly as revealed in the drinks diaries, presumes societally varied everyday routines of ways of drinking water. The UK dominant practices of drinking tapwater as tea and coffee and bottles or cans of soft drinks more than bottled water are almost the polar opposites of Italy, where water is drunk as bottled water, and in much larger per capita quantities, as against tapwater used to make espressos, in a context of very low consumption of soft drinks. Rather than as an independent pole of attraction located in the behaviours of individual consumers and expressed by individual choice, 'demand' is generated by social practices, which in turn can only be understood in the context of societal relational configurations, and their historical emergence.

Finally, the comparative configurational analysis significantly alters the perspective and analysis of the sustainability of bottled water. The simplistic identification of bottled water with tapwater as mutually substitutable, even near identical, makes little sense. Consequently, the view, widely expressed (Gleick, 2010; Royte, 2011; Barlow and Clarke, 2002; Salzman, 2014), that people should stop drinking bottled water because of its higher environmental impact and drink tapwater instead, is misdirected. To be sure, all forms of drinking water, or food for that matter, have an environmental impact, and one of the most significant challenges facing the world today demands their more sustainable production and consumption (Stern, 2014; Harvey, 2014a). In the case of Germany, it is clear that the bottled water drunk is as different from tapwater as any soft drink, and, within the overall consumption pattern, sparkling mineral water marginalises the consumption of soft drinks. So, it is at least arguable that the overall pattern of consumption of water beverages-in-bottles in Germany has a lower aggregated carbon footprint. A similar pattern is seen in Italy, although their tapwater is in all events a less trusted beverage, and there are real and identified health risks in some regions. In the UK, most people drink most tapwater as tea or coffee, both of which have a considerably higher carbon footprint than plain tapwater. Yet, for obvious reasons, no-one is arguing that people should stop drinking tea and only drink tapwater for sustainability reasons. The UK is the converse of Germany, with bottled water if anything marginalised by soft drinks, hence with an overall higher environmental impact but not by virtue of bottled water as such. To understand the sociogenesis of the environmental impact of drinking water, comparative analysis demonstrates the necessity of treating the *overall* consumption patterns of drinking water as plain tapwater, filtered water, water boiled for tea or coffee, bottled water, and soft drinks taken together, rather than any one of them viewed in isolation. Finally, by demonstrating the relationality of consumption within the wider configuration, the challenge of sustainability is firmly placed at a social and political level of configurational change, rather than at the level of moralistic arguments for consumers to change their individual behaviours. The challenge of change for sustainability embraces producers and retailers as much as consumers, including cultures and politics of consumption and production. That might seem to make the challenge greater and more complex. But that is because it is greater and more complex.

4 Divided Delhi

Bricolage water economies and sustainability crises

Aviram Sharma and Mark Harvey

Entering a megacity

The Chief Engineer of the Delhi Jal Board (Delhi's principal public water authority responsible for the network infrastructure) presents a simple but bleak picture. Because growth of the city is unplanned, he argued, the public authority cannot provide a networked infrastructure of water or sewage for a population that, legally, should not be there (Interview, 3rd November, 2011). As a consequence, only 1 in 10 of Delhi's inhabitants, mostly the better-off, is 'connected' to water, and only 1 in 25 to sewage infrastructures.

Likewise, a senior official at the World Bank in Delhi, and an expert on the city's water and sewage provisioning system, said that the World Bank had given up on Delhi, as being 'too complicated to work with' (Interview, 4th November, 2011). The combination of unplanned growth, vested interests, anti-capitalist social movements, and lack of on-the-ground organisational expertise, led the World Bank to withdraw from its earlier attempts to assist in the reform of water provisioning in the city.

Both these counsels of despair might be dismissed as abnegations of responsibility on behalf of organisations deeply implicated in the planning of the capital's water provisioning. Yet they point to a real question as to why it is possible for India to launch a probe to Mars, and impossible to deliver potable water and a sewage system, technologically much more down to earth, to the general population of their own capital. The growth of the city, as with most of the world's megacities including those in India, has been one of chaotic agglomeration at an historically unprecedented scale and pace (Davis, 2007; Bhagat and Mohanty, 2009; Boo, 2012). As we will see when we contrast Mexico City with Delhi, processes of chaotic agglomeration are shaped by historical legacies and by contemporary processes of economic transformation. There are varieties of developmental trajectories in mass urbanisation in the global South. Rural migration and colonisation of urban peripheries and innards creates and continuously destabilises 'settlement', shifting 'archipelagos' of planned provision of public resources, surrounded and invaded by new migrants (Bakker, 2010; Kumar and Aggarwal, 2003), rather than splintering or fragmentation of existing infrastructures (Graham and Marvin, 2001; Kooy and Bakker, 2008). Tapwater and sewerage infrastructures linked to treatment plants with given capacities, are built to connect to relatively stable or at least predictable populations, in housing endowed with plumbing systems, generally along established roadways. Their installation implies planning, yet the social, economic and political environment is characterised by systemic informality (Roy, 2005, 2008), exclusion (Appadurai, 2001), discrimination and disconnection (Anand, 2012). Unlike the motions of urban populations, Mars will still be there when the probe arrives, not a whole lot of different planets in different places.

In Delhi, and similarly in other Indian cities, the very designation of eight different categories of urban spaces as listed in the official Economic Survey of Delhi (Planning Department, 2009, 2013), each with different legal status, both reflects and confirms the process of chaotic agglomeration (Bhan, 2009; McFarlane, 2012): Jhuggi Jhopdi[1] (unauthorised) Clusters, Slum Designated Areas, Unauthorised Colonies, Jhuggi Jhopdi Resettlement Colonies, Rural Villages, Regularised Unauthorised Colonies, Urban Villages, Planned Colonies (Approved) (Dupont, 2004). Not to mention the hundreds of thousands squatting in the streets eluding all official categorisation. Political processes of regularisation, stimulated particularly by elections and the registration of voters, occurs without coordination with urban planning, including water infrastructures, although regularisation also entails rights to these services. Provisioning of tapwater services implies the legal status of urban citizen, which, as we will see, millions of Delhi's inhabitants lack. This is not so much the strategic state deployment of informality described by Roy and others (Roy, 2008; Schwartz et al., 2014), as endemic dis-coordination between different state actors. The process of urbanisation involves the haphazard formation of political, legal, social and economic divisions between the empowered and the disempowered (Chatterjee, 2006), which pattern and condition access to water and sewage.

Much of recent discussion about urban crisis in the provision of domestic water in the global South has been framed in terms of nature of the provider: state, market or various forms of public–private hybrids and partnerships (Marin, 2009). Further, this framing is in turn a response to the lack of water provisioning for huge swathes of the urban and rural populations being attributed to market failure, state failure, and more broadly, governance failure. Markets have proved incapable of spontaneously delivering efficient allocations of resources to meet utility preferences at affordable prices for one of the primary consumer demands. Likewise states – or in the current case – city governmental organisations prove institutionally incapable of allocating resources and planning systems of delivery, through inadequate state structures, corruption, or lack of political will. More broadly, no combination of state, market, and various non-governmental organisations have developed coordinated capacities of governance to ensure delivery of this most basic of amenities. In the words of Bakker (2010):

1 The term Jhuggi Jopdhi has no exact translation, but means shanty dwellings of various kinds.

> Their [urban residents'] experience of water supply suggests that it is simultaneously subject to state, market, and governance failure. Indeed, it is the simultaneous occurrence of these multiple failures that makes water such an unruly resource to manage.
>
> (Bakker, 2010, 47)

At the most common sense level, there is evident failure, most notably in delivering on Millennium Development Goals to supply potable and hygienic water to every individual (however limited the definitions of 'supply' and 'potable' may be) (Gleick, 2004). While interviewing residents of the slums of Delhi, they make it only too clear that water is a dominant concern of their lives, and that they are not getting it. They are anxious to demonstrate just how filthy the water is that they have access to, and the huge efforts they have to go to in order to get it. Moreover, it was equally clear that middle- and upper-class residents also, by and large, had no reliable access to safe drinking water from the public water infrastructures.

Yet the analytical designation of 'failure', whether state, market, governance or any combination of these, is inappropriate and for two reasons. Firstly, failure is only evaluated in relation to an explicit or implicit standard: the perfect market, the ordered and rational state, the fulfilment of a human right or basic human need, perhaps even a standard 'smooth' historical transition or development. Secondly, if water provisioning is placed in the broader and conditioning context of capitalist processes of modernisation and urbanisation – the political, legal, economic and social chaotic agglomeration discussed above – it presupposes that there are 'standards' for these processes as well, by which the emergence of megacities in India might be judged as successes or failures.

Yet, we have seen that the development of the 'municipal hydraulic model' (MHM) in London (Chapter 2) was only the end point of a trajectory marked by sustainability crises, political turbulence, economic disorder and injustices, shifts in organisational scale, and hybrid forms of market and state economies. The route to successful universalisation of the model was strewn with market, state and governance 'failures' to meet its standards – *only if judged retrospectively*. Moreover, this trajectory is one of a capitalist capital city in a particular phase of industrial development, taking a mere 150 years to meet 'the standard MHM'. The technologies of water closets or chlorination were known and in principle available, decades before they became more or less universally standardised. Cesspits, or 'middens', were present in many Northern cities in the UK until the 1960s; in the next step of this socio-technological pathway, outside toilets connected to mains sewage, persisted until the 1970s and 1980s (Glucksmann, 1990, 233; 2000: 138-9). Goubert has demonstrated that France was many decades behind the UK, including in its capital city, in terms of connectivity to both tapwater and sewage (Goubert, 1986). It would be interpreting histories of uneven capitalist development backwards to deem these as instances of failure, whether of markets, state or governance. Moreover, there are contrasting trajectories of urbanisation and 'smoother' institutings of something resembling the MHM:

China has witnessed the emergence of megacities on a scale and rapidity not dissimilar to that of a Delhi or a Mumbai. Yet, under the conditions of a very different political economy, including relatively controlled migration from countryside to city, water and sewerage infrastructures have come much closer, if not close, to the MHM of advanced capitalist economies (Mahadevia et al., 2012). In Beijing, tapwater infrastructures are near-universal, but the quality of water does not fully meet international standards of potability (Browder, 2007). The trajectory of Taiwan, to be analysed later, also approximates to a more 'ordered' instituting of an MHM, with a unified standardised and integrated tapwater system covering most of its territory. But this transition, as we shall see, occurred first under Japanese military occupation, and then under dictatorship, in a command economy contrasting markedly with either London or Delhi, and in very distinctive geopolitical circumstances.

For this reason, we see the emergence of 'economies of water' in Delhi as both a symptom and a tell-tale of a particular trajectory of development and urbanisation. Given the process of chaotic agglomeration, it is beyond our scope to provide a comprehensive analysis of all the heterogeneity and extremes no doubt present in the city. For example, we have not been able to characterise the water economies of street-squatter communities. Instead, we analyse the emergence of two distinct economies of water, the middle-and-upper class mostly connected to tapwater and sewage infrastructures, and the unauthorised slums, commenting briefly on the intermediate economies of water in regularised settlements. Each is characterised by different forms of informality and illegality, as none are reliant on a single dominant source of domestic water for all purposes. We use the term of 'bricolage' economies of water, not in Cleaver's sense of institutional bricolage[2] for natural resource management (Cleaver, 2002, 2012; McFarlane, 2012; Maria, 2008b), but of people/communities making-do with the available haphazard mix of resources unevenly distributed in their different environments, from which they have managed to put together, and regularised, processes of production, distribution, appropriation and consumption of water endowed with varying qualities. In this case, the haphazard mix of resources includes the infrastructures, markets, and housing resulting from chaotic agglomeration, and the capacity to 'put together' an economy of water is conditioned by the economic, social and political divisions that are themselves a characteristic of divided urban 'environments'. In this perspective, it is as much the haphazard and uneven distribution of resources as the practices of the 'bricoleurs' that are formative of bricolage economies of water. These make-do economies of water are thus themselves constitutive of chaotic agglomeration, which is what makes them both symptoms and tell-tales of processes of urbanisation and formation of 'citizenship'.

2 Cleaver describes bricolage as 'the practice of cultural borrowing and adaptation of institutions to multiple purposes and the prevalence of common social principles which foster co-operation (as well as conflict) between different groups of stakeholders'. (Cleaver, 2002, 17).

The megacity in its water environment

Following this initial discussion of how we broadly approach water provisioning in Delhi, the chapter will now first describe in more detail how Delhi has agglomerated, and place this agglomeration in its environmental resource setting, both in terms of rivers and groundwater. We then describe and analyse the two selected economies of water, marking the extent of the inequalities in rights over water resources, both public and private, that distinguish them. In doing so, we deploy the same configurational analysis as used when characterising the successive economies of water in London. We then highlight the various forms of sustainability crises generated by the interaction between these bricolage economies of water and their environmental resources and context. We have seen historical examples of sustainability crises as major triggers for large-scale transformation, and discuss whether Delhi may be confronted with similar conditions that might break out from what now appears an almost fatalistic entrenchment of inequalities, as different economies of water do indeed 'make-do', and in so doing reflect wider potent political and economic divisions of caste, class, and disenfranchisement.

Like most major cities, Delhi is umbilically tied to a major river, in its case the Yamuna, although the urban sprawl spreads from the river plains to the Aravali plateau. Historically most of the population resided on the West bank of the Yamuna, but one feature of the urban agglomeration has been its expansion to the low-lying East Bank, originally considered uninhabitable, but now housing around 23 per cent of the total population of Delhi (Asthana, 2009). The city is dependent on multiple water sources to meet its drinking water needs, both ground and surface, with the Yamuna the primary source for tapwater. The city's water resources are also significantly influenced by the extreme continental climate of Delhi, with three quarters of its annual rainfall delivered by monsoons from July to September, preceded by a hot (40–45C) dry summer from April to June. The city is vulnerable to the cycle of water scarcity and plenty, as the Yamuna rises and falls, including the flood risk to slum dwellers on the East Bank.

Within this waterscape, the whole fabric of water provisioning for the population of Delhi has undergone major historical transformations, and, in contrast to London or Paris, a layering of legacies of two major colonial occupations by the Moghul and British empires. These legacies still shape who gets what water to drink in the Delhi of today.

The period of Mughal rule witnessed the development of their distinctive infrastructures, architectures and aesthetics of water, primarily sourced by the Yamuna river, with a network of canals, dighi (cisterns), nahar (small channels), qanat (underground channels), ponds, wells, and baoli (magnificent step-amphitheatre wells) (Hosagrahar, 2010). The city was much smaller in terms of spread and population, and the majority of the urban population was concentrated in the walled city, around Red Fort. Kahar or bhishti (water carriers) were hired to draw water from the public dighi and deliver it to households (a market for the distribution of water from a public source, an early hybrid). Well water was used for drinking and cooking and neighbourhood wells also served as gathering places, especially for women (Hosagrahar, 2010). The dighi or cisterns, often stone-built and covered, were of a scale akin to Britain's nineteenth-century reservoirs, to ensure water security for fortified towns and cities (Sharma, Y., 2009). The Moghul water regime developed a resilience to Delhi's cycle of water plenty and scarcity, relying on wells and dighi during the dry months, restocked during the monsoons.

Prashad (2001) argued that, due to political uncertainty, wars and the resultant chaos, the extensive Moghul water infrastructure canals systems became neglected and dilapidated during the eighteenth and early nineteenth centuries. It was in this context that British colonial officials transferred the 'modern' piped water supply system, typical of the London of the day, to Delhi. For the first time in 1894 the British ruling elite were provided piped water in 146 private houses located in the Civil Lines and Lothian Road areas (Hosagrahar, 2010). From this initial social division in water provisioning between the colonial and Indian population that continued to rely on Moghul infrastructures, there was a gradual extension of the piped water system throughout the colonial period. Modern water supply extended to 'new' areas inhabited by the colonial occupants on the outskirts of old Delhi.

With the independence of the country in 1947 and partition, Delhi witnessed major inflow of refugees in a rapid, uneven and unplanned population growth (Dutta, Chander and Srivastava, 2005). During the last six decades, Delhi's urban area and population growth has been phenomenal. Between 1961 and 2011, the urbanised area grew from 326 to 1113 square kilometres (Table 4.1). In terms of numbers, it increased from 1.4 million in 1951 to 13 million by 2001 and nearly 17 million by 2011. According to the 2011 census, 97.5 per cent of the Delhi population is now urban (16.3 million) (Planning Department, 2013).[3] London, even during periods of rapid expansion, never faced population change of this rapidity or scale when developing its infrastructures. Although not strictly comparable, London's population doubled in the half century during which its piped water network first developed, whereas Delhi's population increased by twelve times over a similar time span.

The decadal growth rate of the population was attributed to both natural increase in urban population and the phenomenon of in-migration. The major reasons for migration to Delhi are complex: employment, family settlement in Delhi, marriage, education, business, and natural calamities (Planning Department, 2009). Between 1991 and 2001, 62 per cent of the growth of population of Delhi (2.7 million) was due to natural increase and 38 per cent (1.6 million) due to net in-migration (Chakrabarti, 2001). During the last six decades, Delhi has witnessed several waves of strong in-migration, one immediately after Independence (but also witnessed out-migration due to the partition of the country). Secondly, during the post-liberalisation phase in the 1990s poverty-induced in-migration was significant, especially from Uttar Pradesh, Bihar and other neighbouring states such as Haryana and Rajasthan (Chakrabarti, 2001; Planning Department, 2009, 2013).

3 http://www.census2011.co.in/census/state/delhi.html [Accessed 4 October 2014].

Table 4.1 Trend of urbanisation in Delhi: increase in the urban area of Delhi.

S. No.	*Year*	*Urban area*	*Per cent of total area*
1	1961	326.54 sq kms	22%
2	1981	591.90 sq kms	40%
3	1991	700.23 sq kms	47%
4	2001	924.68 sq kms	62%
5	2011	1113.65 sq kms	75.09%

Source: Planning Department, 2009 and 2013

> The squatter population grew by 13.2 percent per annum as compared to the 2.9 percent growth of the non-squatter population [1981–1994]. The integration of these huge mass of people into the city's economic and settlement system has taken place in a complex and dynamic way, largely by the momentum and logic of its own growth, without much intervention of the city's huge but not so effective planning and management bureaucracy.
>
> (Chakrabarti, 2001)

This peculiar pattern continued during the last two decades. The decadal growth rate of the urban population of Delhi is given below (Table 4.2).

Of this urban population explosion, the population of the Central district (the core of the early tapwater infrastructures) now comprises under 5 per cent, whereas the flood plains to the East of the Yamuna, account for 25 per cent. As already noted, Delhi is also divided by settlements of different legal status with differential access to civic amenities (Bhan, 2009). More than half of the city is outside the planning framework. Piped connection was provided to all planned colonies, unauthorised colonies (567) that have been regularised, and urban villages (135). By 2009, piped water and sewerage facilities were also provided to all the 44 Jhuggi Jhopdi resettlement colonies of Delhi. Up to 2008, 1639 unauthorised colonies of Delhi have applied for regularisation, and 1223 were issued provisional regularisation certificates. Most of these unauthorised colonies do not have access to basic civic amenities. Minimum basic amenities were provided to colonies which are settled in private land. In terms of piped water supply, most of these colonies have not been linked from the DJB piped water supply line. According to the 2001 census, more than two million people inhabit JJ clusters. Similarly, slum designated areas are also devoid of basic civic amenities. Based on the pattern of settlement, outside the planned city, different localities in Delhi have been divided into eight categories (Table 4.3).[4]

4 See also Bhan (2009) for further details about the categorisation of settlements.

Table 4.2 Decadal change in Delhi's urban population.

Years	*Total population*	*Urban population in millions*	*Percentage of urban population*	*Annual exponential growth rate*
1951	17.44	1.44	82.40%	7.3
1961	26.58	2.36	88.72%	5.0
1971	40.65	3.65	89.68%	4.4
1981	62.20	5.77	92.73%	4.6
1991	94.20	8.47	89.93%	3.8
2001	138.50	12.91	93.18%	4.2
2011	167.87	16.37	97.50%	1.9

Source: Planning Department, 2009 and 2013

Table 4.3 Type of settlements in Delhi.

S. No	*Type of settlement*	*Estimated population in millions, 2000*	*Per cent of total estimated population*
1	JJ clusters	2.10	14.8
2	Slum designated areas	2.7	19.1
3	Unauthorised colonies	0.7	5.3
4	JJ resettlement colonies	1.8	12.7
5	Rural villages	0.7	5.3
6	Regularised – unauthorised colonies	1.8	12.7
7	Urban villages	0.9	6.4
8	Planned colonies (approved)	3.3	23.7
	Total	13.964	100

Source: Planning Department 2009

According to the 2011 census, there are around 3,340,538 households in Delhi, with an average size of a household of 5.02 (Planning Department, 2013).

The provisioning of water in Delhi

The public supply of drinking water is provided by a state-owned institution, the Delhi Jal Board (DJB). It is responsible for production and distribution of drinking water in Delhi. It is also responsible for collection, treatment, and disposal of wastewater and sewage in the capital. It was constituted in 1998 through an act of the Delhi Legislative Assembly, incorporating the previous Delhi Water Supply and Sewage Disposal Undertaking.[5] The DJB supplies piped water to approximately 2 million domestic consumers inhabiting areas within the Municipal Corporation of Delhi (MCD), the New Delhi Municipal Corporation (NDMC), and the Delhi Cantonment Board (DCB).[6] The DJB also supplies potable water through tankers on demand, primarily to unauthorised colonies and slum areas, but also to supplement water supply to areas connected to piped infrastructures. On a much smaller scale, the DJB also supplies 'packaged' water, '*JAL*', mostly in 20-litre bottles, through small retail outlets, the Jal Suvidha Kendras (Water Convenience Centres). This chapter analyses the water economies operational in the MCD area, which constitutes the majority of the urban area and population of Delhi.

To meet the explosion in demand for water and disposal of sewage, the DJB has been continuously expanding its reach, drawing on surface (river) water from the Yamuna, Bhakhra Dam Reservoir Storage (a distance of 250 kilometres), and the Upper Ganga Canal, as well as from groundwater (Planning Department, 2013). The River Yamuna remains the major drinking water source for the capital. During the last decade, to match the increasing water demand, river water from the Ganga has been diverted through a canal from a megadam located at Tehri in Uttarakhand, around 300 kilometres away from the city. The Ganga water feeds the newly established Sonia Vihar Water Treatment Plant (Asthana, 2009). Correspondingly, with new treatment plants, water treatment capacity has expanded continuously from 66 million gallons per day in 1956 to over 800 million gallons in 2007 (Planning Department, 2009).

The treatment of water sourced from the Yamuna and Ganga has become ever more challenging as those sources have been polluted by sewage flowing directly into it from the city – as the Chief Engineer neatly summed it up, in terms reminiscent of the Thames during the Great Stink: 'We call it (the Yamuna) a sewer not a river' (Interview, 3rd November, 2011). The 48 kilometre stretch of the Yamuna River passing through the city is highly polluted, and even the government agencies accept this fact (Planning Department, 2013; fieldwork interviews). Around 24 major sewage drains in Delhi flow untreated into the Yamuna below the Wazirabad barrage. Further, the waste water and sewerage generated from unplanned localities also flow directly into the Yamuna along open gullies – as we witnessed in fieldwork in the slums. A recent report by Toxics Link (Toxics Link, 2014) measured the levels of pollution from human waste, agriculture and industry in the Yamuna, highlighting the extreme risks to human health from a wide range of contaminants, human, mineral and chemical.

5 http://www.delhi.gov.in/wps/wcm/connect/doit_djb/DJB/Home/About+Us [Accessed 16 May 2011].

6 MCD was a single organisation until 2012, in the same year it was trifurcated into three parts leading to the formation of North Delhi Municipal Corporation, South Delhi Municipal Corporation and East Delhi Municipal Corporation.

Delhi is thus voracious in its upstream claims on resources, and explosive in its downstream environmental impacts. For the Yamuna, there are in principle formal agreements with the four upper-riparian states over the rights to appropriate the water. But, in the dry-season summers especially, with the decrease in water in the upper stretches of the river, there are often conflicts between the DJB and the upper riparian states in Haryana and Uttar Pradesh over the amount of water to be released from the dams in their territory to supply Delhi (Zerah, 1998, 2000). There is no uncontested answer to the public property question: Who owns the flow of the Yamuna? Whether it is the national state, the local states or capital city, the status of 'public good' is unresolved, and different 'publics' contest their rights over control when relative scarcity intensifies competition between them.[7] Equally, the *absence* of control over Delhi's effluent by downstream states on the receiving end of untreated sewage, illustrates the significance of power in the construction of a public good. In IEP terms, the appropriation and degradation of the water resource by Delhi as the dominant political and economic power, and the relative lack of control of both upstream and downstream states, demonstrate how public goods, and who is included or excluded in a 'public', are the outcomes of an ongoing, often conflictual, politico-economic process. The Yamuna never has been a natural 'commons' and is a long way from becoming a politically instituted one today.

Although in quantitative terms only a relatively minor source of the capital's water, groundwater also exhibits similar conflicts and uncertainties over appropriation and control. Firstly, there is a lack of integrated governance over surface and groundwater, in spite of their interactions, with the DJB officially in control of the surface water resources, and the Central Ground Water Board responsible for Delhi's aquifers. Secondly, as we shall see, although groundwater is formally the legal property of the Board, there is no effective control over its abstraction, opening the space for widespread over-abstraction by all kinds of agents – illegal bottled water companies, private water tanker companies, and, last but not least, private domestic households. If there is little control over abstraction, there is even less control over the pollution of groundwater sources.

Three contrasting bricolage economies of water

We have placed the provisioning of water in Delhi in a context of a chaotic agglomeration of a population in an environmental space endowed with water resources whose tentacles spread to the Himalayas, and described the general socio-political context of the state capital municipality and a state-owned water corporation. We now turn to analyse varieties of economies of water that have

7 For example, see http://www.thehindu.com/todays-paper/tp-national/tp-newdelhi/article3534851.ece and also http://articles.timesofindia.indiatimes.com/2012-07-09/delhi/32604068_1_renuka-dam-yamuna-water-water-crisis.; http://www.thehindu.com/news/national/other-states/delhi-haryana-look-towards-centre-to-resolve-tussle-over-river-water/article6134405.ece [Accessed 4 October 2014].

emerged in the city. We cannot claim to exhaust the huge variety of different economic organisation in the city, and instead have chosen to contrast two extremes of a middle-and-upper class planned city and the poor slum economies of water. Then, in between, we briefly analyse an intermediate case of economies of water in regularised settlements, which, at least in part, are the outcome of the tension between the planned provision case and its unplanned and unauthorised neighbouring settlements.

They are all bricolage economies of water. None are characterised by a single, coherent and integrated 'system of provision'. They therefore all enrol combinations of different resources, assembled in large part by the households and consumers themselves, depending on their own material resources, deeply divided socio-economically, and the haphazard presence or absence of infrastructural and natural resources in their immediate environments. In each case, the economy of water is constituted by configurations of production, appropriation, distribution, and consumption: different qualities of water for different populations, in mixtures of formal and informal, legal and illegal rights over water resources. As we will see, informality and outright illegality are at least as manifest in the 'posh' parts of the city as they are in the slums. The analysis of these contrasting economies is summarised using the configurational graphic used as a tool throughout the book.

The middle- and upper-class bricolage economy of water: coping with planning

Despite chaotic agglomeration, the middle and upper classes inhabit parts of the city which are 'planned', in fact, more than that, have witnessed two Master Plans (1962 and 2001), with a third in the pipeline under the authority of the Delhi Development Authority (DDA). These are areas of the city that have been constructed under legal sanction, with planning approval prior to development. They have minimum standards for infrastructure and environment quality, and are lived in by about a quarter of Delhi's population (Planning Department, 2009; Sivam, 2003). During the last two decades, the DDA has developed planned colonies mostly on the outskirts of the city in areas such as Dwarka, Rohini, and Narela. Additionally, private developers have built apartments and colonies in various other locations across the city, such as Shalimar Garden, Sarita Vihar, and Laxmi Nagar. Most of these planned small localities developed by private developers are also middle- and upper-class localities.

Officially, the populations of the planned city are 'connected'. According to the DJB all planned settlements in Delhi are fully covered by piped water supply and sewage pipes. So, in IEP terms, they are supplied by a public body, the DJB, drawing upon the publicly owned water resources (the Yamuna, the Upper Ganga Canal), distributed through publicly owned pipework, connected to domestically owned infrastructures, monitored with meters, and charged with water bills. However, the supply of tapwater is intermittent and unreliable, often only operating for between 1–4 hours a day. Drops in pressure in the pipe infrastructure mean that, although potable when leaving the treatment plants, water becomes polluted from leaks-in, as against leaks-out (Zerah, 1998, 2000, and fieldwork interviews). So, by the time consumers receive the water, its potable quality is no longer assured.

Given these characteristics of tapwater, a range of coping activities have developed directly related to the piped water and to acquiring alternative sources, both in terms of reliability and quality. In relation to tapwater, people have to adjust their daily schedule or chores to the odd timings of its supply – frequently the very early mornings. A more interventionist strategy is the acquisition and use of booster pumps to compensate for low pressure. According to DJB officials, the practice is technically illegal yet has become so widespread that it is now just tolerated. This private domestic misappropriation of public water supplies is, of course, competitive, creating tragedies of the commons if on a minor scale, in terms of whose booster pump, situated where on the chain of supply, acquires water most effectively.

Although possibly a more serious issue in the 'intermediate' bricolage economies of the regularised settlements discussed below, the non-payment of water charges is also a widespread form of private misappropriation of a public good. In overall terms, one DJB official outlined the scale of the issue: '62 per cent of water is non-revenue water. This is shocking.' He estimated that between 600,000 and 700,000 households in planned and authorised settlements receive tapwater but do not pay for it, knowing that they risk little chance of being prosecuted, as confirmed by a DJB legal official. For the maintenance, development and extension of the public piped water system, this lack of revenue flow to match the water flow presents a serious constraint. A vicious spiral of unwillingness to pay for poor quality intermittent supply ensures the continuation of poor quality intermittent supply. Another variety of a tragedy of the commons.

To address the issue of quality and potability of tapwater, apart from boiling, home purification through use of various domestic filtering equipment has become widespread, although also closely tied to income. In a well-to-do middle-class area of Lajpat Nagar, half of the inhabitants used home filtration, as against one tenth in the low-income area of Shahdara II (Zerah, 1998, 2000). Dutta and Verma (2009) found that the usage of home filtration was very high in the planned colonies, for reasons of health, safety and aesthetics of water quality. Depending on the income level people are opting for various home purification technologies in different colonies. In the last decade, there has been a huge growth in the market for domestic filtration equipment, including both UV and reverse osmosis (RO), now the dominant technologies. For every one litre of drinking water produced by RO systems, around three litres of water is wasted (fieldwork interviews). The already burdened water economies are further put under pressure due to these 'technological' interventions stimulated by market forces.

These coping strategies in relation to the inadequacies of tapwater are then supplemented by the addition of two other major water resources: private borewells extracting groundwater, and the commercial market for bottled water, each supplying different qualities of water for different consumer uses. One of the most common components of the middle-class economies of water is the installation of a private tube well, mostly for non-drinking and non-culinary uses, although the water can be processed using the various filtration technologies. Large plastic water tanks are widely used both for borewell and tapwater so as to ensure continuity of supply – the higher the income, the higher the standard of the storage as part of the domestic distribution infrastructure (Maria, 2008a). Depending on location, the use of private borewells has become widespread. As with the booster pumps, private borewells are technically illegal, unless they have the approval of the Central Ground Water Board. It has been estimated that there are between two and three million private borewells, of which only one million have been registered (Maria, 2006). 'Although groundwater abstraction is in theory regulated in Delhi, the authorities in charge of the enforcement of the regulation do not have the practical means of controlling private abstraction. In practice, abstraction is tolerated as a compensation for the failure of the public supply system.' (Maria, 2006, 7). Private borewells are in effect illegal private domestic appropriation of what is legally deemed to be a common public good, the groundwater aquifers. Moreover, this particular route to supplement the inadequacies of tapwater has also resulted in a further distinctive variety of a tragedy of the commons – competitive over-abstraction. This was nicely summarised as follows:

> My house… has only an erratic supply of water. Every house in our area has sunk a borewell, pumps it up to a tank, treats it with RO, and uses it as a complement to piped water. As a result, when we started we drilled to 40 ft, then 60 ft, and now 80 ft, and then people finally gave up. But some have gone down to 120 ft. In those parts of the city where there isn't piped water, there's no option, you can't afford to pay for a tanker supply, which is far too costly. This is taking place in a vacuum, where there is no regulation on extraction. Whether I own or rent a piece of land … there's nothing to stop me drilling for water, if I've got the money. But the water I'm drawing from is not really my water, but everybody's.
>
> (Interview, D Raghunandan, Delhi Science Forum, 7th November, 2011)

In IEP terms, it is important to stress that this form of illegal private appropriation of water is a non-market form of private property – the water is for exclusive domestic use of the householder, rather than traded or exchanged. As such, however, it depends on the non-enforcement of public rights to the water, on the one hand, and a flourishing market in borewell and filtration technologies on the other. The interdependence between private domestic abstraction, processing and consumption of water, the unenforced commons, and commercial markets for equipment constitute distinctively instituted economic processes *in combination*, none of them sufficient on their own. Moreover, the unregulated character of the borewell water means that there is no quality regulation within the private domestic sphere, although arsenic, mineral or chemical contamination presents an unmonitored risk in many parts of Delhi.

The final component of this middle-upper class bricolage economy of water is the dedicated drinking water provided by the commercial bottle industry. Bottled water is supplied mostly in 20-litre plastic bottles for domestic use, and litre or smaller bottles for mobile consumption. Around 64 legal and numerous illegal bottled water manufacturers are operational within the state boundary of Delhi (Sharma, 2014). The branded market is dominated by Bisleri, Kinley (Coca-Cola Co) and Aquafina (Pepsi Co), whereas numerous small legal and illegal firms also occupy a major portion of the bulk water segment. Along with these private players, the public utility (DJB) also supplies bottled water Jar (*JAL*), through several water convenience centres located in different localities. Bottled water manufacturers are solely dependent on groundwater for obtaining the raw water, which goes through multiple water purification processes, finally supplied in bottled jars. The bottled water jars are mostly distributed through small retail shops found in every locality. Based on a small survey taken in five locations in Delhi, Sharma and Bhaduri have argued that perception of risk rather than any normal concept of market consumer choice underpins the purchase and consumption of bottled water. It is frequently consumed when on the move, outside the home by the expanding middle and upper classes. Bottled water, and then only some bottled water, is emerging as the principal trusted source of drinking water amongst high-income groups (Sharma and Bhaduri, 2012).

Overall, the key aspect of the middle-upper class economies of water is the way these different components, legal and illegal, formal and informal, public and market are bricolaged together to form an overall adequate supply of different qualities of water for different purposes. The process of combination, and indeed each component, requires considerable 'consumption work' on the part of householders: installing equipment for abstracting, distributing, storing and processing water domestically, the timely operating of booster pumps, boiling water where necessary, buying and fetching bottled water or organising deliveries, and so on. Given the inadequate quality of tapwater, the final processing of water for different consumption uses occurs in the domestic setting, to a much greater extent than in the standardised municipal hydraulic model. We shall see that Delhi's bricolage economies are distinctive 'instituted economies', quite different, for example from Mexico City, where the dominance of a tanker-cistern-bottled water combination is much more pronounced. The lack of adequate tapwater supplies gives rise to a huge variety of hybrid and complex economies of water in different geographical, environmental and urban settings.

The connection to tapwater in the middle-upper class bricolage economy is itself significant in terms of unequal rights to public resources. Public goods water is the privilege of the more affluent including the rich, and many obtain it even without paying for it. This unequal right to a public resource is then combined with their market purchasing power to acquire the complementary flows of water and its storage and processing: borewells, booster pumps, filtration systems, and storage tanks. Bottled water, the purest market form of water, is also available to those that can afford it. In an upper-middle class shopping mall, Perrier, Evian and San Pellegrino were all selling at more than double the European price.

The configurational figure below (Figure 4.1) highlights the combination of flows of water, processes of production from source through to consumption, the 'changing of hands' and forms of appropriation, the channels of distribution and storage, and the modes of consumption and consumption work. Bricolaged together, and widely reproduced across the 'planned' city, these multiple flows constitute a distinctively instituted economy of water, open to future historical change certainly, but also stabilised in the last decade, and extended in spatial scale across segments of the city.

The poor and lower-class bricolage economies of water

In characterising the bricolage economies of the slums, we primarily focus on slum designated areas and JJ clusters. Together they contain around 34 percent of the overall Delhi population (Planning Department, 2009). Many of these settlements developed on public land and the residents of these localities do not have any legal land holding rights (Banerji, 2005). According to government legislation these settlements are illegal outgrowths and represent the informal spaces in the city. These colonies are inhabited by the lower-middle class and lower classes of Indian society. They are one of the most marginalised and vulnerable populations of the city, primarily rural migrants who have settled in the city over the last few decades. Infrastructures are limited or non-existent, and domestic dwellings are unsuited to normal domestic plumbing. Hence the problem

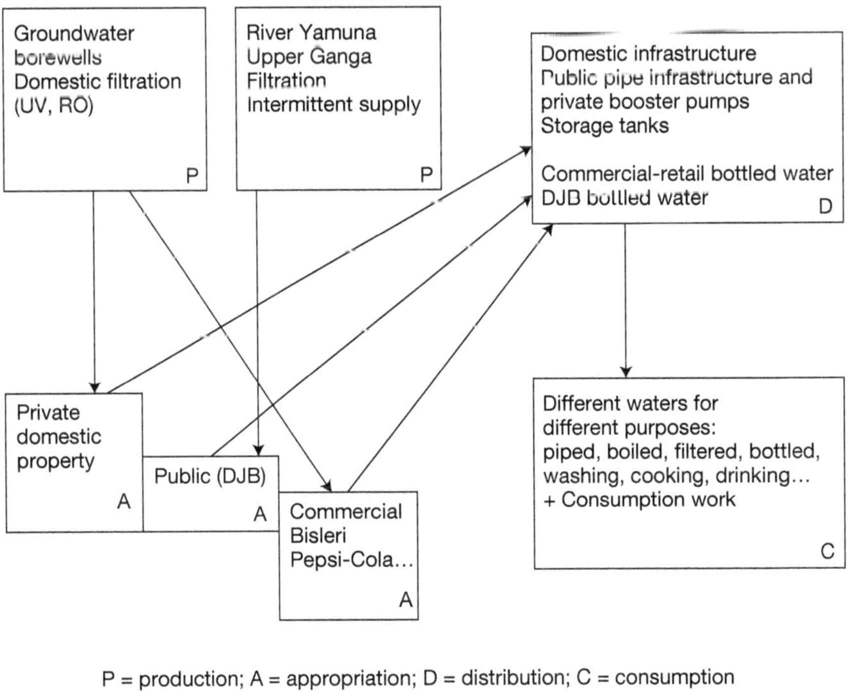

Figure 4.1 The bricolage economy of water in planned colonies.

is much wider than just the provisioning of water and sewerage (Sivam, 2003). Moreover, DJB officials consider connecting illegal settlements to public tapwater as an unacceptable form of legitimisation (interviews with DJB officials), an example of 'strategic informality' and exclusion.

In terms of production and distribution of water in the slums, there are multiple sources, but in a configuration that contrasts sharply with the upper-strata economy of water. For public-goods water, the slums depend on delivery by water tankers and a limited number of DJB standpipes. Alongside these formal economy provisioning systems, there are many informal and illegal complements developed as coping strategies by the citizens of the slums. Llorente and Zerah (2003) state that 80 per cent of the households depend on at least two sources (see also Dutta, Chander and Srivastava, 2005). The water supply through tankers and water drawn from tube wells are often of dubious quality. Moreover, the water drawn from shallow hand pumps or tube wells also provide 'unsafe' water, as the groundwater sources in Delhi are subject to various kinds of pollutions, such as high fluoride, nitrate, higher arsenic, increasing salinity and pesticide concentration (Lalwani et al., 2006; Maria, 2006; Planning Department, 2009). In spite of these risks, lacking the means, the majority of households in the slums consume water without any purification, apart from boiling (Dutta et al., 2005; Shaban, 2008).

In the distribution network of drinking water to these informal colonies, water tankers have emerged as a crucial component over the last two decades. Mostly unconnected to the piped water network, the DJB supplies potable water in these areas by around 1000 departmental and hired water tankers from 25 centres located in different zones in Delhi. The water provided through the tanker is free of cost but is available only on working days. Inhabitants of JJ clusters often report insufficient amounts of water delivery, poor quality and petty corruption charges on DJB tanker water supply (Fieldwork). Turning to our case studies of two slums where we conducted multiple interviews, we point to the main elements of their bricolage economies.

Locality 1 – This locality, established eight to ten years ago, consists of several hundred, mostly one-storey brick buildings, with ramshackle tin roofs, but all connected to the electricity. Alleyways were narrow, but motorbikes and carts could get through in most places. There were large numbers of plastic drums for storing water everywhere outside the dwellings. In terms of the formal economy, water tankers normally deliver water only in the morning, and there was an unmet level of demand for tankers to supply potable water. In addition, young people (mostly male) carry water from a public tap located half a kilometre from the locality, but only between three and four in the early morning or between eight and nine at night, as it was prohibited in the normal day time. In both cases, potable water is a free public good but required a considerable amount of work by consumers, both in acquiring and then storing the water.

Formal water provision, through tankers and collection of drinking water from the nearby public tap however, was inadequate even for basic needs. Consequently, there were also hydrants illegally tapped from the DJB mains supplying the adjoining authorised colony. From that source, they connected two hydrants to feed the entire locality. One of the two DJB hydrants served 24 houses, and a group of mostly men were very anxious to talk about their problem of obtaining water. Each house had the right to take water only once in every seven to ten days, and then could only take 40 litres. The water became available only twice in the day, once in the morning, once in the evening. The pressure was very low, but the water was drinkable according to them. They did not normally boil or purify the water themselves. So, this was a form of collective community appropriation and distribution of scarce water resources, contrasting with the individualised and unauthorised private borewells of the upper-middle class bricolage economy of water.

In addition there was borewell water used for non-potable purposes, and there were several taps from this system, which had its own pipe network in the colony. It was arranged by a local politician, illustrating the patronage-based politics of water operating in the informal spaces of the city. These taps had extremely low pressure, so the outlets were just sticking directly out of the pipe, below ground level, just enough to fill a jug. Again, this scarce water resource was appropriated and distributed by informal collective regulation by the community.

From this account of production-distribution-appropriation of different water resources, it was evident that different types of water served different consumption uses. In this locality, our interviewing attracted small crowds of people, as soon as it was known we were asking about water. For them, it was a dominant concern around which they had to organise their daily lives. They were asking us if we could do anything to help their water situation. They felt themselves powerless to get anything done. The local leader was said to be only concerned about himself, and has made sure that his own house was the best placed for toilets and water.

For sewage, there was no infrastructural provision to the dwellings. For men there were squat toilets open to view, and discharging directly into an open gulley flowing eventually into the Yamuna. Women had to go to public toilets, Sulabh Souchalya, meaning 'easy access toilets'. These were privately run, and charged a couple of rupees (INR) for each visit. The men also have to use it during the rainy season. There were small drains running everywhere throughout the locality which discharged into the large drain.

Locality 2 – The housing was much denser in this slum, with a narrower, even one-person wide, maze of alleyways. Some of the houses had up to three floors, again all connected to electricity. According to one of the respondents, around 15 or 20 years ago there was a big fire in the adjacent Bhumi camp, and many people were killed, especially children. This led the local authority to let them construct brick houses, to replace the tents and plastic and wood structures that existed before. Thus, in an ironic way, tragedy shaped the transformation of this informal space from an extremely temporary arrangement of housing into a more permanent settlement. The population was a mix of Muslim and Hindu, but dominated by the former. As in Locality 1, many were willing to talk, and a group of women animatedly engaged us in their desire to speak their minds about water. All the water and plumbing is outside the homes, nothing inside, so everything has to be carried in cans. Most of the housing units were around eight to ten square yards and had extremely low ceilings. Most of the adults (and, no doubt, children) were working in the informal sector, as maids, drivers, tailors, and small shopkeepers.

As with Locality 1, water tankers were their main source of potable water. To draw water from the tankers, they put pipes into the tanker, and sucked it through so as to fill their jerry cans. Those who did not have pipes drew it directly from the major outlet of the tanker, but they were less able to get any reasonable supply.

They had no, even illegally connected, piped DJB supply connected to standpipes, and relied mostly on groundwater borewells provided by the DJB, but under MCD authorisation. The water was of very poor quality, visibly contaminated to the naked eye – as they were keen to demonstrate to us. There were more than 500 pipes spread within the locality, and the pressure was so low that there were no taps or standpipes. They had to scoop water up from below ground level, using cut-away plastic bottles in order to fill their jerry cans. The outlets were located 10 feet away from each other, but they had to turn the water off at one place, in order to let water come out at the next outlet. Sewage water often seeps into this water source, and so this water is used primarily for bathing and washing clothes. To supplement these inadequate water resources, and to meet the drinking water need, they also used to buy water from the adjacent camp, the Bhumi camp. This camp had an illegal connection from the DJB piped water network, and a group from there used to sell this stolen public water to the residents of Locality 2. They had to pay 60 rupees per month for collection of 20 litres of water per day. So an informal market transaction had emerged for illegally appropriated water from a resource which in origin was a public good.

Here again there were open sewers running throughout the colony. For toilets, they were even more deprived than the first colony, with the 'easy access toilet' being non-gender segregated, with open doors, scarcely ever cleaned, stinking terribly, for which they had to pay 1 INR per visit.

In this locality too, water was one of their biggest problems. But perhaps because it was a long-established slum, the attitude here was quite different from the first. They thought our research could do nothing to help their situation, and they also despaired of any help from politicians. They said India will never change. At the time of elections, politicians used to visit, making promises but doing nothing. They were, however, an important electoral constituency numerically speaking and dependent on politicians for whatever amenities they obtained.

In addition to water resources in the domestic sphere, as with the upper-strata water economies, slum dwellers and lower-middle class populations inhabiting informal spaces also relied, to a small extent, on non-domestic water resources in parallel and in contrast to the bottled water market for the upper strata. A unique water provisioning system in Delhi is the development of push cart trolleys selling drinking water at public places like railway stations, bus stands, market places, and on many busy roadsides. The water vending practice was allowed in Delhi since 1960, and currently there are around 150 licenses granted to sell water by trolleys in Delhi. The number of licenses increased over the years, serving mainly the lower-middle class and poorer strata. All trolleys operating in Delhi are licensed by the MCD, from its 12 zonal headquarters. On average, there are 8–12 water cooling plants operating in each zone, located in industrial or commercial areas (Fieldwork interviews with vendors).

The MCD prescribe a specific shape and size for the water trolley. According to the MCD, the trolley should be four and a half feet in length and two and a half feet in breadth, a standard which had to be met, on penalty of seizure. The water trolleys are authorised to park only on sites permitted by traffic police. In the city limits, the minimum distance between two trolleys should be around 200 metres. These restrictions are applicable only on new licenses, while previous license holders can continue according to their earlier practices. The MCD do not advise water contractors to install any additional water purification technologies, as potable water is supplied by the DJB. However, it is mandatory to write the date of cleaning of the water tankers installed at water cooling plants and the next date of cleaning. For double protection, water contractors were advised to use chlorine tablets. Moreover, owners are liable if someone falls ill after consuming contaminated water from any of their trolleys. If MCD receive any complaints about the refilling of water trolleys from any unauthorised sources, they have power to revoke the license. This water vending system is quite unique, where public DJB water is supplied to contractors for selling on a commercial basis. It is a socially segmented market, posing no competitive threat to the bottled water market oriented to middle and upper classes. The bureaucratic regulation, standardisation and control stands in such stark contrast to the absence of control of other water resources, a single exception to prove the rule. It is prohibited to sell trolley water with slices of lemon, yet all vendors do so, and this alone secures their profit margin (interviews with MCD officials and vendors).

This characterisation of just two slum locations in Delhi of course does not exhaust the variety of slum economies of water, but highlights how economies of water are bricolaged by assembling the haphazard availability of resources in the immediate neighbourhood environments. Slum designated areas in the Old City have some limited connection to piped water (Dutta, Chanda and Srivastava, 2005). In one of the biggest unauthorised settlements of Delhi, Sangham Vihar in South Delhi, private commercial networks have emerged delivering groundwater to households, but the supply is minimal, available for only five hours, and then only on one day in ten (Fieldwork interviews).

To sum up, the slum bricolage economies of water, as with the upper-strata water economy, are comprised of a multiple and complex combination of different forms of production, distribution, appropriation and consumption of water (see Figure 4.2). The contrast is one of two radically different water-worlds. In the slum water economies, the formal public component of water tankers, and, to an extremely limited extent, authorised standpipe borewells, was complemented by various forms of illegal or informal appropriation of groundwater, and in one case a complex illegal market for water between two slums. Both localities were characterised by low pressure, poor quality, and intermittent supply. And in both localities, a dominant feature was the plethora of large plastic jerry cans for storing water in the alleyways, a distinctive feature of the distribution system. The amount of consumer water-work, whether fetching water from distant locations, or dealing

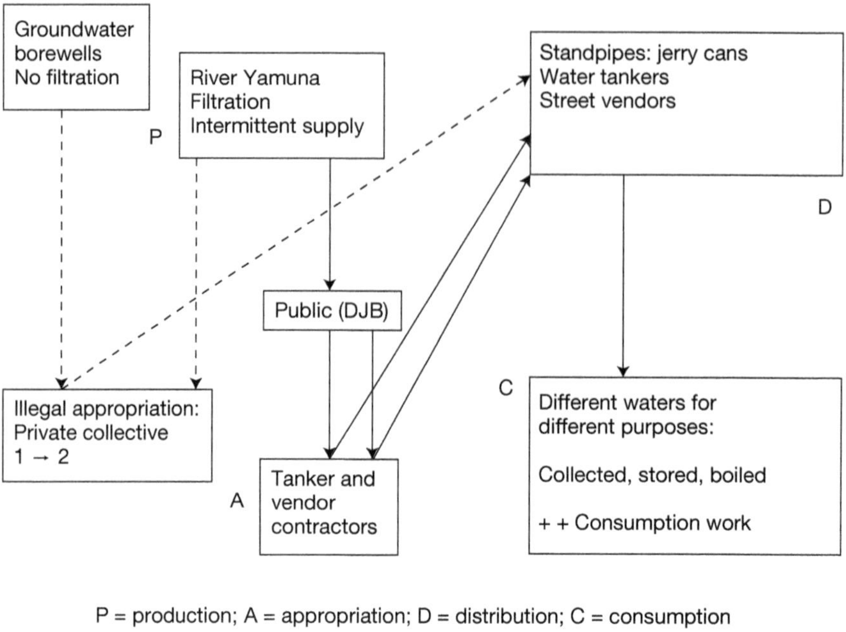

Figure 4.2 The poor and lower-class bricolage economy of water in Delhi's slums.

with it in the domestic setting, was extensive. Considerable resilience and resourcefulness, as well as labour time, was a remarkable feature of this bricolage economy of water. As with the upper-strata water economy, different water resources were used for different consumption purposes, but the difference in quality and availability of water meant that slum water is not the same good as planned urban water. The two slum localities share many features, with slight variations arising from the specificities of location and age of establishment, to an extent that it seems legitimate to characterise them as an example of a multiple replication of locally-bound, relatively small-scale, economies of water across the city. The scale of the slum economy of water is achieved by replication of the basic model in multiple locations, replication achieved by the bricolage activity of people of the slums facing similar daunting challenges to daily life.

Regularising chaotic agglomeration

In this chapter, we have focused on the contrast between the two extremes of middle-upper class and slum bricolage economies of water, in order to illustrate how the process of bricolaging spans the whole socio-economic spectrum. In between these extremes, there has been a gradual process of regularising the unregularised. The settlement categories listed above testify to this process, and include the Urban Villages, the Jhuggi Jhopdi Resettlement Colonies and the Regularised-Unauthorised Colonies. The process of regulation is uneven and strongly affected by political patronage (Roy, 2013; Read, 2012; and fieldwork interviews). Regularisation comes with formal rights to connection to both piped water and sewage infrastructures, and inhabitants are predominantly middle and lower-middle class (Banerji, 2005). As they first emerged without planning, housing infrastructures are poor, and density of population is very high. Often adjacent to unauthorised slums, they are also constantly vulnerable to illegal diversion of water from the piped network, resulting in even less reliable supplies, in terms of both frequency and quality of water, than the middle-upper class residents of the planned city. They are similarly squeezed by over-abstraction of groundwater, so further restricting water from standpipes, which in all events is often non-potable because of its high salinity. Reliance on standpipes and water tankers, both with dubious quality of water, is much higher than in the planned city. Given the limitations of DJB tanker supply, there are also 250 private water suppliers supplying water through tankers in Delhi, especially to these intermediate social strata able to afford payments for tanker water (Daga, 2002; Raghupathi, 2003). Private water tankers are completely unregulated, and deliver untreated and non-potable water. Thus, inhabitants of regularised settlements, with fewer resources than the richer planned city inhabitants, are presented with a different range of resources, each of which is inadequate by itself, with which to assemble their own, distinctive bricolage economies of water, fashioning combinations in between the two extremes highlighted above.

Tolerated division and fatalistic unsustainability

Citizens of Delhi have themselves developed coping strategies in order to meet the need for water, often hooking on to the defective public systems of provision. We have highlighted two sharply contrasting, socially divided, bricolage economies of water, one building on and around defective public piped water, the other on and around a deficient water tanker and standpipe system. These two bricolage economies of water, through the coordination and work of citizens and consumers, provide for water needs at very unequal levels of quality and reliability. The upper strata find much more individualistic and private domestic solutions to water needs, compared to the people of the slums, who, perforce, develop much more collective responses. One of the striking features is that the upper strata have much greater rights of access to public goods water (from the DJB) than the poorer strata, in addition to the rights that derive directly from difference of wealth and income, and their capacity to purchase private commodity solutions, whether that be borewell equipment, filtration technologies, booster pumps or bottled water. They have a double benefit of rights over public and market commodity resources. Furthermore, whereas the richer can purchase their solutions, slum dwellers have a massively greater burden of labour to provide for themselves. They have to work a lot harder for a poorer quality water.

Comparing the two economies of water, it is also striking how both water economies involve illegal and informal appropriation of water resources, whereas it is a common assumption that only the poor, out of necessity, resort to such solutions. Thus, unauthorised borewells, booster pumps and systematic non-payment of water bills, are forms of illegality and informality open only to the rich. People of the slums do not even have the tapwater for which to pay bills. Their forms of illegality involve diverting DJB piped water to shared standpipes, unauthorised collective borewells, or illegal trading of water between two slums.

Another significant dimension of contrast between the two economies of water concerns how economies achieve scale. In the case of the upper-strata bricolage economy, the 2–3 million borewells supplementing their connection to a large-scale piped water infrastructure is achieved through exercising purchasing power in a large and growing commercial market for equipment and filters. Likewise, that purchasing power gives access to national markets for bottled water to supplement their drinking water needs. By contrast, the collective solutions in the slums are small scale and bound by particular constraints of their localities, but nonetheless replicate many of the same features of coping strategies across those localities. Similarly, they complement their drinking water needs by the relatively small-scale operations of street water vendors and a multiplicity of small subcontractors with access to DJB water supplies local to Delhi, a very different market organisation than that of Bisleri, Kinley or Aquafina.

By bricolaging different components together, the different economies of water effectively coordinate production, distribution, appropriation and consumption in different but coherent ways. These economies 'work'. But they 'work' to different degrees: the one providing a reasonable level of 'water well-being', the other struggling to achieve a level of marginal survival. However, the fact that they work almost independently of each other, means that they can also arguably be seen to reinforce and institutionalise social divisions and inequalities of rights to water. As stabilised, self-reproducing economies of water, they may themselves, through inertia or the self-interest of some sections of society, become obstacles to the development of a more sustainable and equitable economy of water. The fatalism of the people in slum Locality 2 gives credence to the possibility that their current deprivation 'may never change'. There is a 'strategic informality' in the governmental tolerance, indeed reinforcement, of these extreme divisions, and of the presence of millions of the capital's inhabitants lacking access to potable water or sewage systems.

But in another, if as yet more latent sense, these economies do not work, at least in terms of long-term sustainability. One of the aspects of Delhi's water highlighted by the combination of a failed public system of provision and the consequent spawning of unregulated bricolage economies of water is the threat to natural water resources, groundwater and surface water, through both over-abstraction and pollution. One consequence of this divided waterscape of Delhi is that a section of the population is much more exposed to waterborne diseases than the other, including gastroenteritis, typhoid and cholera (Dutta et al., 2005). The 'metropolitan hydraulic model' has been much critiqued, but perhaps, in conditions of the chaotic agglomeration of the city, the more fundamental challenge is the failure to develop an integrated commons with universal and equal access, whether in relation to groundwater or to the Indian river systems. The sustainability challenge, the development of public control and coherent rights over the two main water resources at different, but articulated scales, is intimately connected to the current socially divided proliferation of bricolage economies of water, and the pace and scale of urban growth. We noted the division of public responsibilities between different public bodies in Delhi itself, and between the city authorities and the riparian states. But, masked by the formalities of law, the absence of an effective creation of a public commons of either groundwater or surface water is a void encroached upon or despoiled without hindrance by the informal appropriations by privileged individuals or needy communities (Ali, 2003). Informal appropriation of those resources, on the one hand, or, through lack of any effective sewerage system, pollution of both ground and river water, on the other, is becoming an ever more serious threat to the sustainability of all economies of water in the city. This is a *sociogenic* sustainability crisis, peculiar to the interactions between Delhi's economies of water and its natural environment. At present, political indifference and resignation to extreme inequalities of rights to water is combined with a fatalistic acceptance of disease and pollution, running from high officials in the DJB and Municipal Government to inhabitants of slums. We have seen in London's trajectory how sustainability crises were at the same time political and water-economic crises. In this wider sense, the chaotic jungle of competing economies of water in Delhi may eventually lead to a similarly combined political and sustainability crisis, and an historical transformation adequate to the water needs and rights of the population of a twenty-first-century capital city.

Methodological note

Methodologically, our case study was based on over 70 in-depth, semi-structured interviews with a broad range of the principal actors in the water economies of Delhi conducted over two years between 2011–2012: senior executives and officials in the Delhi Jal Board (DJB – public utility) and Municipal Council of Delhi (MCD); a Trades Union official at the DJB; senior representatives of three NGOs involved in campaigns for water provisioning; a World Bank water expert; a marketing director of a major commercial bottled water company; municipal bottled water retailers; tanker water contractors; street water vendors; and citizens of two unauthorised colonies (slums), and three regularised settlements and urbanised 'rural villages'. Interviews in the slums and regularised settlements were often group interviews, with people collecting around the interviewers, anxious to explain their situation with regard to access to water and toilets. In addition, we analysed government statistics of water supply to various parts of the city.

5 Mexico City

Up in the sky without a river

Daniel Figueroa and Mark Harvey

In London and Delhi the historical evolution of the city and capital appeared utterly dependent on a switch from groundwater to river water as a condition for urban growth, a sufficient volume of water flowing through the heart of each city to supply a high density population eventually numbering millions. As major cities, they are unexceptional in this respect. Mexico City is hence a global anomaly: at 2,200 metres above sea level, it has no river flowing through it, and now accommodates a population similar to Delhi's and over twice as large as London's. Consequently, it has a quite distinctive society–nature dynamic, facing challenges of a different order in the formation of its economies of water. Equally, as we shall see, what comes in must go out, and so it also has economies of waste water, both from rainfall and from domestically and industrially used water. It doesn't risk creating a river-sewer flowing directly under the noses of parliaments of historical London, or contemporary Delhi. Its wastewater problem is perhaps the greater for that. In the words of an expert engineer involved in the construction of wastewater infrastructures, which discharge into the peripheral river basins, well outside the city:

> 'Currently the river is crap, it's shit. It's full of things you can't even imagine. Everything from the city comes out through there [...] There are no flora or fauna, all the fish are gone, all the animals.'
>
> (Ricardo Morayta, Ingenieros Civiles Asociados, interview, 23 July 2013)

Words reminiscent of those describing the Thames in London in the mid-nineteenth century, and the Yumana in Delhi today, yet referring to a reality not directly under Mexico City's nose.

Resources and rights

If there was ever a case for the social construction of water scarcity, the actions of the Spanish conquerors of Mexico might lay claim to it. In 1521, while taking possession of the Aztec capital, Tenochtitlan, constructed on an island in a valley basin lake surrounded by high mountains, they drained the lake and many of a chain of lakes of which it was part (Sosa, 2010a). The Aztecs had built a complex system of dams, partly to separate brackish water from rainwater, and partly to control the level of the lake in response to the extreme variation of seasonal rainfall. During the siege of the capital, the Spanish destroyed the dams, never to be rebuilt, and left a contradictory historical legacy of scarcity and abundance: a future city both deprived of rainwater as a source of supply, and subject to periodic flooding, often severe. But, of course, the city would never have grown so big had the lakes not been drained in the first place.

The contemporary city is therefore built in a valley basin, and has grown, chaotically, into a complex of urban centres, administratively divided into sixteen districts (delegaciones), with a total population of 21 million. The historical legacy at first left the city dependent on deep wells as the principal source of water (with 618 in the metropolitan area alone, INEGI, 2011a). As the population grew in the twentieth century, many of the aquifers became exhausted or contaminated (Conagua, 2011).

So, as with other major cities, there was indeed a significant switch from ground to surface water to supply the growing population, but, given its location, this meant the construction of dams at a distance, and the transportation or pumping of water up to 2,200 metres above sea level. Although the switch was only partial, the city first became dependent on the Lerma dam reservoirs in the 1940s, and, when this became increasingly inadequate for the growing population, the Cutzamala dam system, one of the largest in the world, became a major source of water. The Lerma water source had been pumped into high-level reservoirs, then feeding the City by gravity. The Cutzamala system, with its seven dams, pumps water from over 100 kilometres distance, creating a demand for electricity equivalent to the total requirements of a city of 1.5 million people (Tortajada, 2003, 2006; Delgado-Ramos, 2014). During this period, the population of the city grew from 1.7 million in 1940 to 8 million in 1970, 15 million in 1990, to its present population of 21 million. Currently, the Mexico City conurbation still obtains 73 per cent of its water from local aquifers, supplemented by 18 per cent from the Cutzamala system, and 6 per cent from the Lerma river basin, with a further 3 per cent from nearer rivers and springs (Conagua, 2011).

Unlike the abstraction of groundwater from wells from beneath the city, the Lerma and Cutzamala dam systems have been a continual source of conflict over rights to water resources. The stresses placed on water resources from the Lerma and Cutzamala river basins by Mexico City exemplify the tensions between Mexico's water resources being deemed a national public good, belonging to the whole people, and the competing demands by different social groups for different demands, domestic, agricultural and industrial. Water had been enshrined as national public property in the constitution of 1917, but successive reforms of governmental control and management, combined with weaknesses in enforcement of that control, exposed the emptiness and incoherence of these formal constitutional rights. The Lerma River runs for 714 kilometres from Mexico's high central plateau providing resources to six states along its route to Lake Chapala (Mestre, 2001). It also provides water for Mexico's second largest city, Guadalajara. In 1970, the downstream populations suffered drought while water was being pumped upstream to supply Mexico City, provoking social conflicts along its course (Cohen and Reynoso, 2006). Between 1990 and 2003, Lake Chapala lost 80 per cent of its volume, causing major social and political unrest (Vargas, 2008).

Likewise, major conflicts for rights over water resources have characterised the development of the Cutzamala system to this day (Kloster and Alba, 2007). The Zapatista Army of Mazahua Women in Defence of Water protested both against the displacement of people caused by the construction of dams, and by their dispossession of rights to water consequent upon the diversion of supplies to Mexico City (C.L. de la Peña, 2011; Gomez, 2010). Protests were taken before the Latin America Water Tribunal, which adjudged the Cutzamala system as causing a widespread degradation of water resources, damaging the rights over water resources for people affected by it (F. Peña, 2004). The *Agua Rios Y Pueblos* activists laid a superior human rights claim to water above those enshrined in national rights to water in the constitution.[1] Dr Anahi Copitzy (Colegio de Jalisco), a Mazahua herself, emphasised this tension:

> I don't see anything positive [from the Cutzamala system] for the ethnic groups. Firstly, it's about imposition. No one asked them, no one asked them "Do you want the system to go through here and take your water?". I'm not saying they had rights over the water, because it's property of the nation, but they used that water. So there is no consultation with the ethnic communities and it's a system of imposition, antidemocratic and it belongs to another time.
>
> (Interview, Anahi Copitzy, 10 July 2013)

Bakker has highlighted the complexity of overlying rights between communities and populations at different social and geographic scales to the water resources that transect their spaces, pitting, in this case, rights to water for the urban poor (and rich) in Mexico City against the rural populations downstream (Bakker, 2007, 2010). Moreover, many of the capital's inhabitants had migrated from those downstream rural communities. So, in spite of the formal national constitutional position, there is no clear answer to the question of who effectively appropriates, or 'owns', the rivers Lerma and Cutzamala. There are conflicting and overlapping rights to control over the resources, between national ministries (SEMARNAT[2]), regional administrative hydrological regions (the Regiones Hidrológico-Administrativas), municipal and city authorities (e.g. SACM[3]), riparian ethnic communities, industrial and mining enterprises, and farmers (Mestre, 2001; Tortajada, 2006). Appropriation is a mix of formal and informal, legal and a- or il-legal, control of finite resources. In this sense, it can be characterised as a tragedy of the anti-commons (Heller, 1998): too many social entities with competing and inconsistent property rights for many given resources; the mirror

1 (http://www.aguariosypueblos.org/) [Accessed 4 October 2014].

2 Secretaría de Medio Ambiente y Recursos Naturales.

3 Sistema de Aguas de la Ciudad de México).

image of a tragedy of the commons, too many people unrestrictedly competing for the same property rights over a single given resource. Critically, moreover, these competing rights arise in, and are hence conditioned by, the specific natural environmental context of providing water to a riverless sky-borne Mexico City from river resources transecting many different social and political spaces.

The Spanish historical legacy has had as much impact on waste- and wasted water: the challenge of disposing both of high levels of rainfall from May to October, and of the volume of sewage generated by a population of 21 million. As with early nineteenth-century London although with very different effects, the tragic 'solution' has been to combine surface water with sewage in an infrastructure indiscriminately pumping huge volumes of both into the river basins below (Interview, Felix Hernández Gamundi, Director, Grupo Mundi, July 2013). The city itself has been flooded frequently, with major health risks from the mixing of sewage with rainwater.

> The major risk in Mexico City is sewage. Not because of coverage, coverage is reasonable. As the city sinks, some areas are more affected than others. [...] we have to pump wastewater out of the city and from an energy perspective this is madness. New tunnels are being constructed to carry out wastewater. However with floods you have wastewater flooding the city.
>
> (Hugo Contreras, Industrias de Agua, interview, 2 July 2013)

Already at the turn of the twentieth century, the Gran Canal was constructed to carry the mixed sewage and rainfall water out of the city, and in mid-century this infrastructure was supplemented by a new Deep Drainage infrastructure, after the development of illegal slum dwellings and the general growth of the city's population. In sharp contrast to Delhi, however, contemporary Mexico City has a high level of domestic connection to sewage infrastructures, with over 98 per cent of dwellings. However, only a very small percentage of this water, perhaps as low as 10 per cent, is treated by the city authorities, resulting in untreated sewage flowing directly into the river systems (Sosa, 2010b). In 2010, the 112 water treatment plants in the metropolitan area were capable of treating only a small proportion of the city's total effluent (INEGI, 2011a; Conagua, 2011). Consequently, there is now a project to construct a huge treatment plant at Atotonilco, due to be finished in 2014, which will reduce the level of pollution to the river system. Unlike contemporary London, however, such distant water treatment plants preclude any reuse of waste water.

Like nineteenth-century London, however, pumping human effluent into a river system has a habit of coming back to bite you. Wastewater has its predominate effect on downstream rural populations, with risks to health, especially associated with the use of the wastewater for agricultural production. But then the fruits and vegetables also return to the city, and are a major source of gastro-intestinal disease. As we shall see in the next section, contaminated water also affects the domestically available tap and tanker water, and, in ways quite distinctive to Mexico, results in chemical disinfectants being added to water in the home for preparing food.

> You will find that the locals [in downstream rural communities] sell cauliflowers the size of watermelons ... The food that is grown with wastewater has a lot of health issues, for the people that grow it and for the people that eat it.
>
> (Interview, Cecilia Tortajada, Third World Centre for Water Management, 8 July 2013)

The flows and returns of water and wastewater in Mexico City and its surrounding environments are thus conditioned in quite distinctive ways compared with Delhi and historical London. In contemporary Delhi there are conflicts with riparian states upstream for sources of water, and with downstream ones for contamination of the river Yamuna. But these are characteristics of a capital with a major river flowing directly through it, and the nature of the conflicts are tied to that particular society–nature interaction. In both cases, there are issues of competing positive claims over, and negative impacts on, finite water resources. Yet the interactions between their contrasting political economies of water and their natural resource environments endow the conflicts over these issues with quite different dynamics. And it is to the economies of water within Mexico City that we now turn.

Economies of water in the capital

The previous section established the general context for the development of the varied provision of water to domestic households in the city, and contributes to explaining their distinctiveness. Mexico City could not be like a Delhi or London, although much more than this contextual difference is required to explain its peculiarities, making it the bottled water/soft drink capital of the world. Three distinctive economies of water will be identified as the predominant ones characteristic of the city, especially Mexico City itself, as against the Mexico City Metropolitan Area (MCMA). They are distinguished primarily by virtue of the differences in access to tapwater (continuous, irregular, none), which also corresponds roughly to social divisions of wealth. But, as tapwater is generally regarded as non-potable, and indeed largely below the standard required for preparation of food, it would be misleading to see relative provision or deprivation of tapwater as a mark of deprivation of water overall. The tapwater in Mexico City is not the tapwater of London or advanced western economies. All the capital's economies of water rely on two other principal sources of provision: bottled water for drinking, and tanker water as either the dominant or the supplemental supply of non-potable water. So, each of the economies of water is characterised by a particular combination of two or three of these components: tap, tanker and bottled. In this respect, Mexico City stands in sharp contrast to Delhi, in that each of these three components has a city-scale presence, with tapwater being the only one that is less than universal. There are different waters for different purposes, different provision-consumption relations in configurational terms. For that reason, each of these components will first be presented in turn, before turning to how different configurational combinations characterise the social divisions and inequalities of economies of water within the city.

The tapwater system of provision

As already indicated, tapwater is generally of a low quality, according to several recent studies (Mazari-Hiriart et al., 2005; Sosa, 2012). Even according to official accounts from the national organisation responsible for overseeing water provision and the hydrological regions, 29 per cent of Mexico City and the metropolitan area is polluted, with a further 54 per cent highly polluted (Conagua, 2011). Sosa established a Water Quality Management Index, demonstrating the widespread failure to produce potable quality water, and the failure of chlorination treatment to prevent high concentrations of biological pathogens (Sosa, 2012). As Roberto Constantino, a water economics expert from the Universidad Autonóma Metropolitana, explained:

> Sacmex (SACM) will tell you 'drinking water meets the regulations', and yes they do. However, the official norms come from the 70s, they are not up to date and the purification process relies mainly on chlorination. Forgive me, but who in their right minds drink water from the tap? [...] The major risk of the city is water quality. It's a time bomb.
>
> (Interview, 18 July 2013)

Up to the early 1980s, people with access to tapwater generally boiled it for drinking and cooking, and so, with the additional 'consumption work', prepared it for final use. However, the 1985 earthquake damaged the pipe infrastructure, leading to widespread leakages, loss of pressure, and hence contamination. There was a cholera pandemic in 1991, followed by successive outbreaks during that decade (Mazari-Hiriart et al., 2005; Borroto, 2000). As we shall also see in Taiwan, the fear of cholera is a powerful and long-lasting influence on consumer trust in tapwater, and effectively from that time, tapwater widely ceased to be a source of water for drinking, or for much food preparation.

> Public entities have given us poor water quality. People mistrust the government and we end up buying bottled water. People are intelligent, they might get sick and they end up buying water quality. Even government officials go to a conference or a meeting and they all have bottled water.
>
> (Agustín Breña, water engineer and hydrologist, Universidad Autonóma Metropolitana interview, 23 July 2013)

Even in Mexico City (Federal District), where 99 per cent of households are actually connected to the piped water infrastructure, only 62 per cent of households have continuous supply, and in the poorer areas, less than 50 per cent (UNAM, 2010). The overall picture for the whole of the MCMA is one where intermittent supply is widespread, reflecting deep social divisions in access to water, far from a universal good, whether public or market. As we have seen in Delhi, intermittency of supply is intimately connected to lack of water quality security, and, as we shall see below, underpins the dominant characteristic of domestic water infrastructures.

If rights of access to tapwater are unevenly distributed, the mode of provision is complex and conflicted. On the one hand, Conagua is the national organisation with responsibility for overseeing the delivery of water to Mexico City, the management of the Lerma and Cutzamala systems, and the Regional Hydrological Administrations. It is ultimately responsible to the national Ministry for the Environment and Natural Resources (SEMARNAT). On the other, Conagua is responsible to the Mexico City government department, SACM (Sistema de Aguas de la Ciudad de México), which is responsible for the urban pipewater and sewerage infrastructures, under Article 16 of the Federal District Water Law (Ley de Aguas del Distrito Federal). SACM in turn has delegated management of the water system to four private corporations, as concessionaries. Before going into further detail on these concessionaries and the extent and nature of this 'privatisation', the conflicted nature of the public responsibility and control over water provisioning points to an absence of a unified 'public' governance, an absence of an overall representative public agent. The national organisation, Conagua, and the City administration, SACM, have respectively responsibility to deliver water to the city and to manage it within the city, and, given the flows of water across these governance scales, inevitably conflicts arise. At one point, Mexico State, for example, demanded payment for water that crossed its territory from Mexico City. More frequently, when water shortages arise, or irregularities of supply deteriorate, Conagua routinely post notices assigning responsibility to officials in SACM. As one of the concessionary private management company managers explained in relation to the district for which he is responsible:

> There is a coordination issue between Conagua and SACM. For SACM it is very easy to say that there is no water in Iztapalapa because Conagua hasn't provided them with water. This is because they don't want to assume responsibility for this. Since there is no regulator, they confuse the people. On the other side Conagua might also say they provided enough water when maybe it isn't true. This would be very easy to avoid with a third party [regulator] but there is none.
>
> (Hugo Contreras, Industrias de Agua, interview, 2 July 2013)

However, in addition to conflicts between national and city governance scales, SACM itself has a distinctive limitation to its powers over water provisioning within the city: it does not control its own budget, responsibility for which is assumed by the City Treasury.

> They [SACM] have no assets and no revenue. The treasury approves their budget and their expenses are around 10 billion or 12 billion pesos and income from water bills is around 5 billion. There is an imbalance. Is the director of SACM worried about this? No because it doesn't matter how much income they receive, their budget depends on the treasury and will remain the same regardless of revenue.
>
> (Hugo Contreras, Industrias de Agua, interview, 2 July 2013)

The separation of budgetary from management responsibility for the public provision of water, and indeed the substantial subsidy for water provisioning from general taxation, not only fractures the public character of tapwater as a politico-economic good, but obscures and partially disconnects the flows of money and water, of which Swyngedouw speaks (Swyngedouw, 2004). However, this governance–budgetary disconnect is no 'decommodification' of water, let alone socialisation, or deliberate political provision of free or below cost water. And, to make the public character of tapwater even more turbid, SACM is subject to political pressures from the City government to vary the price of water in different constituencies, to appease their voters.

> We have a very high subsidy. It's a political issue. In the marginalized areas, and speaking specifically about Iztapalapa, they have around 2 million people. It's like another city that you wouldn't find in other places of the world or here in Mexico. Unfortunately in that zone the aquifers have very low quality so what they do is they just give them water for free. They are charged a very low figure for water.
>
> (Agustín Breña, water engineer and hydrologist, Universidad Autonóma Metropolitana, interview, 23 July 2013)

Where water supplies are irregular, or inadequate, particularly in the poorer districts, tapwater (and, as described below, public tanker water) is priced very low, or at zero. The budget deficit borne by the City Treasury is therefore not so much related to either the free provision of a public good, or even a deliberate and transparent redistributive subsidy from general taxation to poorer districts lacking adequate water supply, as an expression of fractured governance and political clientelism. One of the primary underlying aspects of this local political pricing of water is the recognition of resistance to paying for an unpredictable, variable and low quality supply. Yet, the consequent disconnection and lack of transparency between the charges for, and cost of, water no doubt contributes to the lack of investment in developing the infrastructures and addressing the social divisions in access to supply. These all are testament to the challenges of instituting an economy of water as a universal public good even within a Mexico City Metropolitan Area scale.

In the context of this disconnect between flows of money and flows of water, the privatisation of the management of tapwater takes on a Mexico City flavour, in a way that reinforces Bakker's insistence on distinguishing between privatisation and marketisation or commodification of water (Bakker, 2003a, 2010). From 1993, SACM partially devolved the management of water provisioning to four companies, three of which were owned by French multinationals (Veolia and Suez). Initially, the contracts were for a fixed term of ten years, but they were extended first to 2011, and have now almost assumed the character of permanent incumbency. The management concessions, however, were highly limited, only delegating the reading of meters, the issuing of bills, the recovery of payments, and the servicing and maintenance of the infrastructures (SACM, 2012). As already evident, they have no role in either the fixing of prices or in capital investment. Even the water meters themselves are now purchased and provided by SACM, leaving the private concessions only to install the meters. Moreover, given the political sensitivity of privatisation, SACM has gone to great lengths to make the private companies invisible, so that most people think they are dealing with SACM when paying their bills, and the offices where people pay the bills are displayed as belonging to SACM. Hugo Contreras of Industrias de Agua (owned by Suez) again:

> Mexico City government wanted the private contractors to be invisible. A lot of people don't even know we exist although that's irrelevant. Our role is just to serve the customers, to charge them and solve questions or problems.
>
> (Hugo Contreras, interview, 2 July 2013)

The economic organisation of the tapwater provider side is thus characterised by a public–private combination certainly, but predominantly under the fractured control of various public organisations, lacking any transparent democratic control of a unified public. Given the limited, subordinate and almost invisible nature of the private management concessions, the public character of tapwater as an economic good is more bureaucratic than democratic, more clientelist than collective. But before leaving the tapwater component of Mexico City's water economies, the provider side of the system of provision is incomplete without considering the central and complementary importance of domestic infrastructures. Given the insecurity of supply, most households have installed ground-level cisterns and roof-level tanks, fed by water pumps – depending on the wealth and resources of the household. In other words, domestic infrastructures play a much more strategic role in ensuring continuity of supply than in the advanced tapwater economies of the North.[4] Moreover, as we shall see later, these infrastructures are complementary to both tap and tanker provision systems, both of which, in their own ways, deliver supplies intermittently.

> SACM provides you a partial system. You, as a user, have to build a cistern to guarantee continuous provision, you have to get a water tank, you have to buy a water pump. You have to spend money on a cistern, on a water tank, on an electric pump, on bottled water. With this I am telling you that they provide a partial service and the population compensate at a micro level to secure provision and quality. So you see it in your house, you buy a filter and you spend money but millions of people are doing that. So the system does not solve 100% this issue, the rest is solved at the micro level. This is also another

4 As we have seen, the interconnection between private-domestic and public provision was historically contentious, and there has been a constant evolution of the respective responsibility of the private-domestic and water provider. In the event of tenancies, there is another modal variation in the interconnection, depending on whether the landlord is public or private. In the case of the former, the public domain covers both domestic and external infrastructures.

> reason why people don't like paying or when the tariff goes up. They ask "why am I paying you this if I am spending money?" and with all of these things they sometimes do not have water.
>
> (Arsenio González, Expert, Urban Planning and Development, Universidad Nacional Autonóma de México, interview, 24 July 2013)

Again in marked contrast to Delhi, these cistern-roof tank-water pump infrastructures are a very widespread private-domestic property counterpart to public–private water providers. They achieve a significant economic scale, even greater than the public infrastructures of tapwater. These domestic infrastructures are predicated on the typical domestic buildings, with detached houses dominating the urban landscape. Mexico City is sprawling. In the metropolitan zone, 79 per cent of all housing consists of detached buildings, only 14 per cent are flats, with tenement housing accounting for just 5 per cent (INEGI, 2011a, b). In Mexico City, in these dwellings, 54 per cent overall have cisterns and 78 per cent water tanks (2.4 million households); while in the metropolitan zone the figures are 38 per cent and 67 per cent respectively (3.7 million households). In some of the boroughs, no doubt as a reflection of wealth, households with cisterns rise to 79 per cent, water tanks to 88 per cent, whereas in poorer areas it can fall as low as 18 per cent and 69 per cent respectively (INEGI, 2011b). So some households can afford much greater water security and flexibility than others.

However, although these domestic infrastructures enhance water security to an extent, they certainly have the opposite implication for tapwater safety, such storing of already contaminated water in hot climates only intensifying risks to health. By delivering water unreliably and unfit for drinking or food preparation, the piped infrastructures therefore combine with two other channels of water provision, one complementary to drinking (the bottled provision system), the other in relation to security (the tanker system), and it is to the latter that we first turn.

The tanker system

As Agustín Breña (interview, 23 July 2013) affirmed, relying on tankers for the bulk of non-potable water has become the norm in Mexico City, along with many other urban centres in Mexico. There are both public tankers directly supplied by SACM, and private enterprise tankers. If the settlement is a legal one, the citizens have a right to public water tankers, and pay only a small administrative fee (about £4). But, partly as a result of the bureaucratic and lengthy delays in delivery of up to six weeks, many inhabitants also depend on private water tanker contractors, when they pay a price per volume. The contrast between the water tanker system in Delhi's slums and the Mexico City system could not be greater – in spite of their apparent similarities in transporting water by tanker. In Mexico City, people order deliveries by the tanker-load of water to fill their cisterns, with tankers varying in size from 5,000 litres to 20,000 litres. There are some surface similarities with the Ecuadorian city of Guayaquil's *tanqueros*, which, however, are all private, mostly very small-scale operators, supplying 35 per cent of the population, complemented by street vendors, often illegal, selling water at higher prices to the poorest households (Swyngedouw, 2004). Yet they clearly are inserted in very different overall economic configurations of water, conditioned by the contrasting polity–economy–nature interactions of a port city at sea level and a city in a high-mountain valley basin, providing a context for distinctive trajectories of historical development.

It is a household delivery system, with only limited opportunities to share tanker-loads with neighbours. For both the private and public tanker system, consumers take a laminated map of their exact house location to the SACM or contractor's office when ordering the tanker, and the driver returns it on delivery. For private tankers, the fee for water in a 5,000 litre tanker is 0.16 pesos per litre (roughly 7 pence), reducing to 0.14 pesos (roughly 6p) from a 20,000 litre tanker.

The district township of San Miguel Xicalco is typical of a legal settlement, where inhabitants rely heavily on the tanker system. The water deliveries can take place more or less at any time of the day or night (fieldwork interview, July 2013). Their first option is to obtain public water tankers but, especially in the drought months of December to May, they may be forced to pay for private tankers, often from distant parts of the city. In this township, the official statistics may show that households are connected to the piped infrastructures: connected they may be, tapwater they may have none. *Agua Potable* (drinking water) is emblazened on the sides of many of the tankers. As both private and public tankers rely on SACM water for their source, and as the water has not passed through the pipe infrastructure, the water may indeed be drinkable in the tanker. However, the water is not for drinking as soon as it is discharged into the household cistern, which is also the receptacle for tapwater, and may frequently mix with it. So, although the tanker system is on a city-wide scale and provides a necessary complement to water security to the piped system, the people of Mexico City have to look elsewhere for water fit for drinking.

One of the consequences of the shared infrastructures between tap and tanker water has induced a final process of 'qualification' of water to make it fit for use in food preparation and cooking quite distinctive to Mexico. Consumers purchase a disinfectant, with several brands, such as Microdyn, widely available in supermarkets and grocery stores. The disinfectants contain electrolytically dissolved silver, and vegetables and fruits contaminated by Mexico City's sewage waste are then regularly disinfected using contaminated tanker or tapwater with the disinfectant additive. The practice is very widespread – as is the market for the disinfectant products. Extra consumption work is necessary to prepare food ready to eat, in the context of a particular configuration of water and food provision!

Bottled water

As a consequence of the tap and tanker qualities of water, Mexico City inhabitants, in common with most Mexican citizens, rely heavily on bottled water, essentially for drinking but also for some food preparation uses. The average annual per capita consumption is 248 litres of bottled water, and a further 166 litres per capita of soft drinks (Castano, 2012), a combined total of well over a litre a day.[5] So, again in contrast to Delhi, there is widespread consumption of bottled water, not just confined to the middle and upper classes. Moreover, it is also overwhelmingly sold in bulk, in 10 to 20 litre bottles.

The bottled water market is dominated by large corporations, notably Coca Cola (FEMSA), Pepsi Bottling Group, Nestlé and Danone. In 2013, Danone had 26 per cent of the market, Coca Cola 23 per cent and Pepsi, 13 per cent. However, there are also many small contractors, employing ten people or less (INEGI, 2012). All draw their water from the public SACM source, and then process it industrially using reverse osmosis (RO), ultra violet, and carbon filtration. The companies pay commercial rates for the public water. In IEP terms, therefore, the water is a public-private good, with the market players acting as intermediaries between a public source and end consumers. The price that the bottled water providers pay is a public monopoly fixed price (Conagua, 2012), in contrast to the price that consumers pay, which is indeed a competitive market price. The price per litre of bottled water varies from 2.5 pesos to 3.7 pesos, plus the initial cost of the bottle of 50–70 pesos. As a very rough guide, companies pay .0014p per litre to SACM for tapwater, and once processed and distributed, sell it at 3.75p per litre in a 20 litre bottle. This compares with the price paid for privately operated, non-potable, tanker water of 0.7p to 0.8p per litre.

However, as with tap or tanker water, bottled water is an incomplete provisioning system, without the addition of domestic infrastructures and consumption work to finalise the water for consumption. Given the widespread dependency on bottled water (over 94 per cent of Mexican consumers, on a daily basis, Barquera et al., 2008), home deliveries are widespread, and often include a contract to supply a water dispenser and cooler. Many households, however, buy their own distinctive and decorative bottles to store and dispense their bottled water, often combined with water coolers. Consumers coordinate and organise deliveries, or indeed, buy bottled water from the myriad of retail outlets, and transport it, often in trolleys, back home. Finally, many people complete the qualification process by flavouring the water with the common fruit-flavoured powders such as Tang, or, of course, making coffee or tea with it. Cooling, boiling and flavouring thus complete a process of qualification of water beginning with the filtration and chlorination of SACM public water and continuing through the industrial processing undertaken by bottled water providers. Mineral and natural spring water are quite marginal to the consumption practices of Mexico City.

Bottled water in Mexico City (including soft-drink and flavoured versions) is significant above all as the dominant channel for delivering drinking water. Although undoubtedly conditioned by affordability (the evidence for which is

5 Mexican consumption of bottled water is thus nearly 10 times greater than the UK, 2 ½ times greater than Germany, and 15 times more than Taiwan, combined with substantial multiples for the consumption of soft drinks.

difficult to demonstrate in the absence of reliable statistics), commodified drinking water constitutes the specialised drinking counterpart to the other two channels, both of which deliver non-potable water, even water so inferior as to be no longer domestically treated by boiling or filtration. Recognising bottled water as the main drinking water channel is thus significant as we now turn to consider three Mexico City economies of water, distinguished more by variability in access to non-potable water through the other two channels. It is possible, of course, to consider the emergence of a specialised, commercial, bottled water channel as a response to a failure of either tap or tanker provisioning systems, whether public, private, or public-private, to deliver the all-purpose drinkable water, as the assumed global standard to which advanced Northern economies have become accustomed. From a more neutral, anthropological perspective, these economies of water can be explained in the historical perspective of a distinctive evolutionary trajectory, in a particular environmental context, going back to the Spanish legacy, the impacts of earthquakes, and the biology of cholera.

Three economies of water

Given the formation of a more or less metropolitan area scale for the three main channels of water, it is possible to represent a general configuration of production, appropriation, distribution and consumption, and then to use this as the point of reference for describing each of the three economies of water characteristic of contemporary Mexico City.

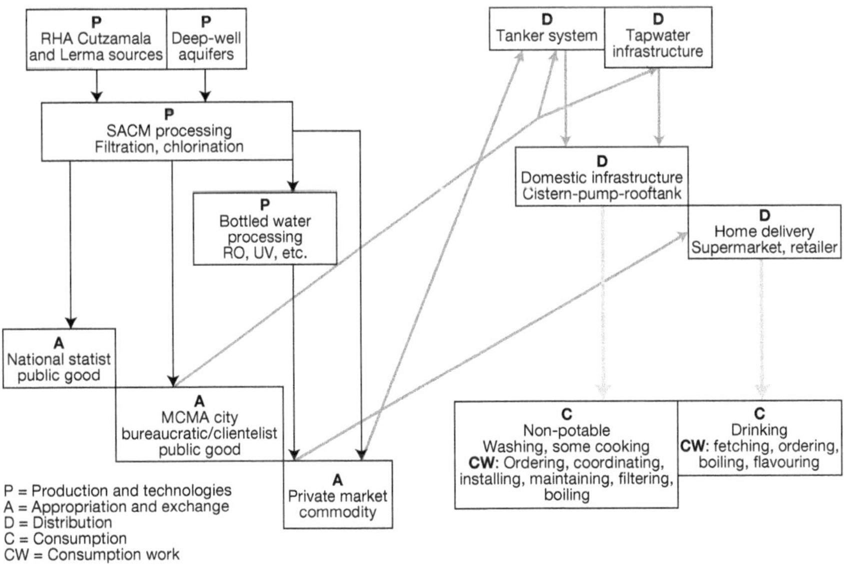

Figure 5.1 Economy of water in Mexico City Metropolitan Area.

The relational configuration schematises the flows of water in space, through productive (qualitative) transformations, successive appropriations, ending up with consumption and the work of consumers in preparing for its final use. Figure 5.1 describes the field of possibilities for inhabitants of the MCMA in obtaining water for the various purposes of washing (personal, clothes, etc.), toilets, cooking and drinking. It illustrates the flows of water through the system of provision to the end consumer, analysed below. The principal significant variation between the three economies of water is determined by the extent to which households are dependent on tap or tanker water.

The dominant economy of water

Most households in the MCMA and even the Federal District of Mexico City itself are connected to the mains tapwater infrastructure, but only receive intermittent supply (UNAM, 2010), including the 80 per cent who receive some water daily. Consequently, in terms of relative volumes of non-potable water, households may receive as much as 80 per cent of their water from tankers, both private and public, even though officially connected to piped infrastructures (Gonzalez interview, July 2013, and fieldwork interviews with consumers). In fieldwork in the township of San Miguel Xicalco, it was found that households officially connected to the piped network received no water at all, and were totally reliant on tanker water. This economy of water is then also characterised by the high levels of bottled water consumption for drinking. This economy of water characterises a very broad social mixture, and, at the lower end of poverty, may well rely on boiling water for drinking, and reduced levels of bottled water consumption. Similarly, they may well depend more on public tanker water, and receive lower volumes of water overall, than those that can afford private water tankers, and on a more frequent basis.

The middle and upper strata economy of water

As with Delhi, there is a paradoxical greater right of access to public tapwater supplies for the middle and upper classes living in the central areas of the Federal District. These households are assured of continuous supply of non-potable water, at low cost. Reliance on tankers is much reduced, and it can safely be assumed that there are few constraints on the amount of bottled water that is purchased, or indeed the use of water coolers and dispensers, which typify middle- and upper-class homes.

Lower class and illegal settlement economies of water

At the opposite extreme, there are many households, 37 per cent in the MCMA as compared with less than 1 per cent in the Federal District, which are not connected to the piped network. In lower-class legal settlements and especially in illegal settlements, there is thus a total reliance on tanker water. Moreover, in the negative mirror image of the rich, those living in illegal settlements are not entitled to public tanker provision so have to pay the higher commercial rates for their water. The poorest thus are totally dependent on market-commodity water, whether tanker or bottled, for their water supplies. Again, although there are no reliable statistics, it can be assumed that consumption of bottled water drinks is limited by affordability, with a concomitant higher exposure to health risks or to the costs of boiling water for consumption. Moreover, given that domestic infrastructures are also market goods, there are no doubt corresponding inequalities in the security and safety of these across the whole spectrum of the population of the MCMA.

Undoubtedly, this characterisation of the three economies of water in the MCMA simplifies and reduces a wider variety of effective rights to water, and the social divisions related to them. Political clientelism is certainly a source of variation between the 16 delegaciones. Equally, the growth of the city has been chaotic, and, as with Delhi, provisioning systems are slow to respond if they do at all. Nonetheless the generic configuration presented in Figure 5.1 does capture the large-scale features of Mexico City's waterscape. The IEP approach reinforces and develops Bakker's insistence on the need to escape any neat dichotomies between public and private/market economies of water. Indeed, as a relational configuration, it highlights the multiple interdependencies between public and market actors. Bottled water producers, even major multinational corporations, purchase their source water from a public undertaking – SACM, as do the much smaller-scale water tanker operators. There is much less dependency on private appropriation of natural water resources (although these too are national state property), from springs or deep aquifers, partly because of their high levels of contamination in the Mexico City valley basin. So, even though Mexico City stands out in contrast to many major cities for its dependency on the commercial bottled water market for the provision of drinking water, if one views the overall flows of water and changes in appropriation, water passes from national state property, to city government property, before entering the sphere of the market commodity, and terminating its ownership journey in domestic water cooler infrastructures under private-domestic property rights.

The subordinate role of the concessionary corporations for the management of tapwater supply has already been noted. As has been frequently remarked, monopolistic system of distribution embedded in an integrated urban tapwater system supports, in this case, tapwater as a public bureaucratic good. The final pricing of water, the transfer of ownership to the domestic consumer, however, permits much greater fragmentation, with political clientelism resulting in widely varying prices across the city. This contrasts with Taiwan, where water as a national public good under a single state corporation, has a uniform price across the whole country (Chapter 6), or with London where such variation of pricing between the boroughs eventually led to municipalisation and price formula standardisation at a particular geographical scale (varying in relation to property values).

Yet, Mexico City and the MCMA is distinctively reliant for the bulk distribution of water (non-potable) on tankers, a system that is much more open to small contractors, even if many might establish quasi-monopolies for particular urban districts. These small private contractors, although dependent on public good water for their source, illustrate how different distributive technologies open up a space for different patterns of appropriation and market formation, even when SACM, distributing water virtually free, remains the dominant single player in tanker distribution of water.

The particular form of tanker distribution of bulk water in Mexico City moreover can only have co-evolved with domestic infrastructures as a generalised feature of a built environment dominated by detached houses: the cistern-pump-roof tank technology. Deliveries of 5,000–20,000 litres at a time presuppose investment by householders, and thereby institute a particular interdependency between domestic-privately owned infrastructures with the tanker water owners, whether public or private. The administrative cost or pricing of water linked to the bulk exchange of water is hence also embedded in this technology–property–economy nexus.

Pursuing this configurational analysis, the household investment in the cistern-pump-roof tank technologies also presuppose a co-evolution with a distinctive form of consumption work related to them. The whole relational complex depends on consumers arranging the installation, servicing and maintenance of the system, planning and organising the ordering of supplies, preparing a laminated map for the tanker driver, and then doing whatever needs to be done to the water for one of its potential uses, 'requalifying' to make it fit for purpose. Although unpaid and informal, such activity is critical to the economic organisation of the tanker system: they work to make it work. What Mexico City citizens do to make this system work is quite different from what Delhi slum citizens do to make their system work: children and women filling up jerry cans, getting up at 5.00 am in the morning, generally on a daily basis, and so on.

Thus by drawing out the particular relational interdependencies between Mexico City's organisation of forms of production of non-potable and potable water, the different proprietorial rights and exchanges over infrastructures and flows, the technologies of distribution, and the forms of consumption, we can see them as instituted with more or less historical stability, over different urban scales. Moreover, the configurations outlined above in the three economies of water and the generic schema (Figure 5.1) represent only a moment in an ongoing process of change. To understand some of the key dynamics of that change, we must place these politico-economies in their environmental context, because, without doubt, crises in sustainability already manifest suggest that these instituted economies are far from stable.

The sociogenesis of Mexico City's sustainability crises

Not all sustainability crises are sociogenic, in the sense of being generated by human activity. The earthquake of 1985 fractured much of the piped infrastructure, resulting in a high level of leakages which wasted up to 50 per cent of water pumped up to the City (Tortajada, 2003). Yet even such 'naturogenic' events clearly have distinctive impacts on particular societal organisations – piped infrastructures vulnerable in areas of high seismic activity standing as an example. Failure to repair the system after 1985, and the resulting contamination of tapwater supplies, was seen as a significant turning point in the capital's economies of water, as noted above.

However, the naturogenic disruption of the mains infrastructure has been compounded by the distinctively sociogenic effects of overabstraction of groundwater, leading to major subsidence in many urban areas. From the 1930s, the growing population drove the increasing abstraction of groundwater (Sosa, 2012), until the city had sunk to an alarming extent already by the 1950s, prompting the suspension of use of several deep wells. Today, Conagua estimate that four of the fourteen aquifers available to the city are already overexploited (Conagua, 2011). According to Miguel Ricaño (SACM) and Roberto Constantino (Universidad Autonóma Metropolitana (Interviews, 17 and 18 July 2013) subsidence remains a persistent and significant problem for maintaining the integrity of the tapwater infrastructure.

Thus overabstraction of groundwater creates subsidence, which then leads to increased waste of water through leakages, so placing even greater reliance on obtaining sources of water outside the City. There is a vicious spiral in which the sustainability crises of groundwater displaces demand onto the Cutzamala and Lerma systems, which in turn are prone to shortages and increasing demand from agricultural, industrial and domestic uses downstream. Already, the Cutzamala system has met limits in its storage capacity, partly exacerbated by long-term changes in rainfall. As Miguel Ricaño (Director, Institutional Development, SACM, Interview, 17 July 2013) insisted:

> The Cutzamala system was not created for drinking water for Mexico City. The system was adapted to seven dams that provide water to the whole Basin of Mexico. [...] The major challenge is the problem of storage in the Cutzamala system. Every year there is less rain so we have to cut the service.

The view of the SACM Director is echoed by Dr Anahi Copitzy:

> The [Cutzamala] system is environmentally unsustainable. They keep taking water from other basins, such as Michoacán and Mexico State. [...] Two or three weeks ago people in Cutzamala, Guerrero, were without water for their households and their lands. ...Why would you take water to another region and leave the original place unprotected? [...] there is a constant process of impoverishment.
>
> (Interview, Dr Anahi Copitzy, Colegio de Jalisco, 10 July 2013)

But there are further twists to this vicious spiral. Leakages from subsidence or earthquakes result in both intermittent supply and contamination of tapwater. That in turn has meant that tankers became the main bulk distributors of non-potable water on the one hand, and that tapwater was no longer relied on for drinking water, on the other. So SACM processed water, pumped up to the city at a huge energy cost, is then transported around the city either in tankers or bottles, with further energy impacts. Different component processes of sociogenic sustainability crises have become interlocked with each other. The contamination from a leaking sewerage system outflowing directly into the downstream river system, only belatedly and partially treated, closes a circle of inflow and outflow, contributing yet further to the complexity of Mexico City's polity–economy–nature interactions. And then, the rains come. The flooding of the city further stirs the mix of inflows and outflows (Tortajada, interview 8 July 2013). We are back in the legacy of the Spanish destruction of the Aztec dams, and the contradictory legacy of scarcity and abundance. Roberto Constantino draws out the paradox:

> The greatest problem in terms of water in Mexico City is not the lack of water. We have a surplus. That's why the greatest concern over the last years is floods and clearly there are no floods unless we have a surplus of water. We have a surplus. There are two contemporary problems: not enough underground water to provide the city with water, that's why the city is sinking, why Xochimilco is sinking. The second problem is that the city continues to flood. It's normal, the city is a drainage basin. It's an artificial valley and we live in the lowest part of the valley.
>
> (Interview, July 18 2013)

So, in its present trajectory of development, Mexico City's economies of water are locked in to a continuing spiral of sustainability crises: growth in demand from legal and illegal settlements, overabstraction of groundwater, subsidence and wastage through leakage, intensified reliance on external river resources such as the Cutzamala, wastage of rainwater, flooding and pollution of river and groundwater (Delgado-Ramos, 2014). Placing the politico-economies of water back into a maelstrom of environment interactions may seem too obvious an analytical move, given that springs, rivers and rain are almost taken as symbolic of the natural. The Aztec hydraulic state, sometimes nostalgically portrayed as a more balanced economy–nature order, would have prevented the emergence of a megacity. Yet, centuries later it has been historically superseded by politically conflicted, mal-governed, socially divisive, disjointed economies of water which, equally remarkably, provide essential life-support for the society and polity of 21 million people in Mexico's high mountain valley metropolis. At least, for the time being.

6 Taiwan water

Little island, waves of power

William D.H. Li and Mark Harvey

Why Taiwan? Taiwan is a small island (population 23 million, 13,807 square miles, compared to the UK, 69 million population, 94,500 square miles). Yet, precisely because of its size, Taiwan is on the one hand more analytically 'manageable', and on the other, Taiwan has been subject to big waves and powerful natural forces. The island's water and its historical development have to be understood in terms of the external political forces that have shaped it. Taiwan has long been a pressure point in the geopolitics of the region: first, as victim of the Japan–China conflict and military occupation by Japan in the late nineteenth century, and then, post-Second World War, as a strategic domino in the Cold War tensions between the US and China. Water infrastructures were built, modified, and developed in significant part as a consequence of direct intervention by these external forces. New economies of water emerged.

In turn, these politico-economic forces are complemented by the environmental stresses of a subtropical climate: tropical diseases (especially cholera) and typhoons have swept over the island many times and to devastating effect. These natural disasters have also triggered major and successive changes in Taiwan's economies of water. So Taiwan makes it possible to develop an analysis that brings together these dimensions of international political and natural forces.

And yet, Taiwan is not a spot in the Pacific Ocean barren of its own proper internal dynamics of growth, including its social and political transformations, and indeed its own traditional cultures of consumption, notably drinking boiled water and tea (now bottled and iced). So small islands, and their water histories, can ask big analytical questions concerning the dynamic relations between geopolitical and natural forces, and thus stretch our conceptual understanding of the complex interactions between them. In short, Taiwan allows us to introduce and develop new dimensions of analysis for understanding the simple act of the drinking of water.

The chapter will be divided into two main parts. The first part will trace the development of urban tapwater and village infrastructures, through the Japanese period of occupation until the Second World War; the post war period, when the military dictatorship of the Kuomintang (KMT) was in power; and the eventual democratisation of Taiwan in the 1990s. Through most of the period, there was a marked urban–rural divide in the island's economies of water, but nonetheless a gradual process of scale formation in both these contexts. Scale formation of economies of water are critical for examining what is meant by the concept of 'public' in public good, firstly in terms of urban versus rural, with very different rights over water of quite different qualities, and different organisations of exchange and pricing, in these contrasting contexts; secondly, in terms of local, provincial or national public economies of water, and consequently, whether the people of Taiwan universally enjoy the same rights to water under a unified economy of water. The key aspect of the Taiwanese experience is that this evolution culminated in a substantially integrated national economy of tapwater, more so than, for example, the current regionalised and privatised economic organisation in the UK (as discussed in Chapter 2). Most Taiwanese citizens now drink similar water, pay the same price, to a single state corporation, the Taiwanese Water Corporation, across most of the island. In analysing this evolution, the interactions between the geopolitical and national forces with the environmental challenges of diseases and typhoons will be an important focus of analysis.

Nonetheless, despite being a strong centralised 'developmental' state, the core national economy of water has evolved unevenly. So, in the second part of the chapter, we zoom in on the South West of the island, to the city of Kaohsiung and the neighbouring Pingtung County. Here, in the local context of successive crises of water provisioning and governmental failure to fully extend the core national infrastructural skeleton of tapwater, a quite Taiwan-distinctive formation of a water economy has emerged with the creation of 'drinking-water garages'. They look just like petrol garages, only drinking water, not petrol, comes out of the nozzles of the pumps. For a non-Taiwanese stranger, at a distance you could not tell the difference between a petrol and a water garage. This more local focus, moreover, reveals processes of qualification of water, also quite distinctive to Taiwan, with differentiated qualities of filtered water selling at different prices in an extensive private market for drinking water. As we'll see, however, the peculiarities do not stop there. These drinking water markets take as their source either public tapwater, or groundwater under license from nationally owned springs. So processes of qualification in the production of varieties of filtered tap or spring water are articulated with distinctive combinations of public–market appropriation, distribution and exchange.

The making of a national 'good'

The emergence of the modern drinking water system: the Japanese era (1895–1945)

The modern phenomenon of water supply from piped networks was initiated at the end of the nineteenth century, at the very beginning of Japanese colonial rule. In 1898, the first modern drinking system was set up under Japanese rule in Tamsui, a port next to Taipei (Department of Health, 1995: 78). Following the First Sino-Japanese War (1894–95), Taiwan was ceded by China to Japan at the signing of the Treaty of Shimonoseki. The construction of drinking water facilities became one of the major priorities of the government's agenda during the period of Japanese occupation, laying the foundations of the public drinking water system. However, urban modernisation was certainly not the only, perhaps not even the main, objective of the construction of the 'municipal hydraulic model'. The spread of contagious diseases posed a serious threat to Japanese rule over Taiwan at that time, as cholera and the plague were endemic in the region. When the Japanese troops first landed on the nearby islands of Penghu to take over Taiwan after the First Sino-Japanese War, about one third of the 6,194 soldiers were affected by cholera and other contagious diseases, and 1,247 of them died due to the illnesses they contracted during the first 50 days of their arrival on Taiwan. More soldiers died from disease than were killed during the battles with Taiwanese rebels during the period of the takeover (Department of Health 1995: 160). A Japanese army doctor wrote down a detailed description of his shock when encountering the poor living conditions in the Taiwanese settlements in 1895:

> The street is filthy. Human and animal excreta along with dirty water are all over the muddy streets in which pigs are fed on the street. The wooden huts are without windows and stink inside. There is no sewage or waste water system and human excreta flood elsewhere outside the huts. Women use water from streams and ponds for the households' drinking water and for washing clothes. In large cities such as Taipei and Tainan, the drains along the roads are full of stagnant waste water, and stink and are disgusting.
>
> (Dun, 2005: 13–14)

His account found echoes in the more official record of the high death rates from tropical diseases in the US Health Report of 1902:

> JAPAN – In Formosa, during the period from January 1 to January 21, there occurred 100 fresh plague cases, of which 80 ended in death. From January 15 to January 21, there were registered in the port of Tamsui alone, 13 plague cases, all ending in death. According to official statistics, the total number of cases of plague in 1901 amounted to 4,519, of which 3,634 terminated in death.
>
> (US government, 1902, p.831)

Thus disease control and hygiene dominated the provision of urban tapwater during this period, rather than notions of public goods, or indeed rights to water of any quality. Gotō Shinpei, who took part in the construction of the sewage and water systems in Tokyo in the 1880s, became the health advisor to the first Governor of Taiwan, Kodama, in 1896. He later became the first civilian governor of the island in 1898, and laid the foundation for the state's regulation of the drinking water and sewage system. The initial Taiwan drinking water plan was established by William Kinnimond Burton (1856–1899), who served as the Japanese Colonial Government hygiene consultant and had previously been the Resident Engineer at the London Sanitary Protection Association before being invited to Japan in 1887.

The colonial drive to control disease did not stop at providing hygienic water infrastructures. It extended to imposing new disciplines of hygiene within the household: sanitary inspectors and hygiene 'police' examined the state of people's homes in order to ensure that hygienic practices of cleanliness and order complemented the investment in infrastructures especially during the Gotō Shinpei period (Chin, 1998).

For those not covered by urban tapwater infrastructures, in particular the rural population, shallow wells were the typical and traditional means of providing drinking water at that time. Some were shared among locals and provided by local collective organisations such as temples. Other households had wells next to their houses, although these could be also shared with neighbouring households. These household economies of water, therefore, involved considerable amounts of work and time, carrying and storing, boiling and cooking to provide for household needs. Although generalised amongst rural populations across the island, the rural scaling was one of replication of independent, household-neighbourhood systems of water provisioning.

However, if the rural domestic economies of water were in one sense largely untouched by Japanese occupation, in another sense, Japanese rule entailed bringing the water resources of the island overall under state control. The island's rivers and mountain areas were effectively appropriated by the state. Water storage facilities such as reservoirs and dams were built at that time. A government land survey was first conducted in 1895, and those landowners who were deemed not to hold land documents such as title deeds, had no right to the land. Huge amounts of land became state-owned. For those mountain areas where the major water sources were located, land and rivers that had been communally owned became government-owned and controlled. Kin and communal rights over mountain resources for food and water, hunting and gathering, were outside any money economy, let alone private property or individual human rights. Securing control over water resources and the construction of dams was a significant aspect of Japan's genocidal war against indigenous mountain people, traces of which are still visible today. Control over mountain rivers was certainly an 'appropriation by dispossession', in this case, by the colonial state, a 'nationalisation' of communal rights. By contrast, prior to the Japanese invasion under the Ching dynasty, much of the island fell outside effective government control, except in the urban areas. Any water ducts for drinking and irrigation for public use which had been privately built before Japanese rule, were also nationalised and controlled by the colonial government according to the 1901 Act related to Taiwan's public use of water springs and ducts (Chen, 2001).

In an important sense, therefore, the development of urban infrastructures, sourced by mountain rivers and dams, can be seen as an instrument of colonial rule, providing benefits to the urban population certainly, but also imposing Japanese control and order over a new economy of water and natural resources; over the ordinary routines of domestic behaviour of the Taiwanese urban population; and, last but not least, over the biological and natural environment of their occupation.

By 1917, about 16 water systems had been established to provide drinking water for the major urban areas in Taiwan such as Taipei, Taichung and Kaohsiung. Adopting Western technology in a way typical of the Meiji restoration, sand filter-beds, chlorination, pumps, piping and water meters for the households were constructed and installed to provide drinking water to the urban public. By 1945, the publicly supplied water coverage had been raised from nil at the time of the Japanese occupation to over 17 percent of total population in Taiwan (see Table 6.1; Wang, 2001, p.1061, Table 5-11-1-4; Bank of Taiwan Economic Research Department, 1965).

This Japanese phase of Taiwan's evolving tapwater economies and infrastructures are placed in the longer-term perspective of later developments in Table 6.1, showing the extent of coverage up to the end of the Second World War in 1945. The Table also provides a point of reference for the subsequent trajectory of subsequent developments, to which we now turn.

Industrialisation Era: scaling up in town and country, 1945–1973

When the Nationalist Chinese government took over the island after the Second World War in 1945, much funding was required to restore the tapwater systems island-wide due to the damage caused by the war, and the number of water systems had been reduced from 123 in 1940 to 88 in 1950. The government lacked funding to develop drinking water and priority was given to channelling available funds to other basic infrastructures, such as electricity, transportation and industrial development. Fewer resources were therefore available for the construction of water systems in the early 1950s, which resulted in the provision of water developing very slowly.

Yet, given Taiwan's strategic significance for the USA especially with regard to Communist China in the period following the Second World War, along with the Mutual Defence Treaty between the US and the Republic of China (1955–1980), significant aid was provided to embrace Taiwan within US global hegemony. New water systems were a major beneficiary, with 61 percent of the total investment of NT$ 18.7 billion in water infrastructures coming from foreign aid (Wang, 2001, Volume 5). The main international support for water construction came from the US Aid and especially loans from the International Development Association of the World Bank (IDA). This massive injection of aid raised water standards in Taiwan well above the standards prevalent in other, less geopolitically strategic, developing countries. It is a distinctive Taiwanese connection between water and power, not exactly a Marshall Plan, but certainly an incorporation into the Western sphere of influence just offshore of China (Barrett and Whyte, 1982).

However, as in the pre-war period of Japanese occupation, geopolitics combined with bio-environmental forces. In the 1950s, drinking water became a much more serious concern as huge numbers of soldiers and refugees coming to Taiwan from mainland China had transformed the urban landscape. Squatting and shanty towns spread around the peripheries of towns and cities, posing major health risks by the

Table 6.1 Public water systems and its coverage in Taiwan, 1905–2010.

		Coverage %	*Total population (million)*	*Urban population/ total population %*	*No. of tapwater systems*	*No. of simplified systems built*
The Japanese occupation era	1905	1.4	3.02	3.82	2	
	1910	5.6	3.2	4.94	6	
	1915	8.1	3.4	4.97	18	
	1920	9.8	3.6	6.97	22	
	1925	13.1			45	
	1930	16.9	4.56	11.27	72	
	1935	10.9			91	
	1940	14.6	5.89(1939)	14.9(1939)	123	
	1945	17.89			88	
The first industrialisation phase	1950	26.77		27.61(1951)	88	77
	1955	27.52	7.87(1951)		116	
	1960	29.67		34.07	132	11
	1965	38.32	10.79		133	100
	1970	43.26	14.68	45.11	132	330
The national integration phase	1975	58.5	16.14		246	270
	1980	78.6	17.80	59.16	211	96
	1985	85			222	
	1990	90.1	20.35	66.44	232	
	1995	90+				
	2000	90+	22.21	67.99		
	2005	90+				
	2010	90				

Source: Tapwater: Taiwan Water Association 2006a, p.44–100.
Note: Statistics do not include the Taipei city area. Urban population: Population of over 20,000 in urban settlements, from Huang, 2002. Simplified water systems: Department of Health, 1995, Vol.2, p.765, Vol.5, p.1456.

1950s. More than one third of the population in the city of Taipei were squatters (Li, 1998). A new governmental public health agency, an environmental sanitation laboratory, was set up at the Provincial Government level in 1955 and was responsible for public health care that covered drinking water, human waste and the disposal of garbage (Department of Health, 1995, p.1446). Since the war, over 2,000 people had died of cholera before another outbreak of cholera in 1962 resulted in a further 24 mortalities (Department of Health, 1995, p.384). This was a turning point for the further extension of public drinking water, even though the coverage rate for drinking water had by 1960 increased to encompass 30 percent of the overall population. In particular, these persistent threats to human health from cholera finally shifted emphasis towards providing safe water and sewer facilities for the rural population, not only urban citizens.

Consequently, during the period of industrialisation up to 1970, whereas the number of urban tapwater infrastructures added only 11 to their pre-war total, 518 rural water infrastructures, supplying tapwater to villages, were created. These were known as 'simplified water systems'. They consisted of borewells, filters, pumps, a chlorination technology, and a piped network to the village houses. To ensure safe water, they were standardised systems, and so, although small scale, their replication across the island provided an extensive scale by the proliferation of small units. Simplified village-based water systems played a critical role in public water provision in the 1960s (see Table 6.1). Foreign aid-assisted water systems had been introduced in 1950 and 77 simplified water systems were set up by the US Taiwan International Partnership Agency referred to as the *Sino-American Joint Commission on Rural Reconstruction (JCRR) (*中國復興聯合委員會). These simplified systems gradually replaced the ad hoc, individual borewells, with unregulated, hygienically insecure, water systems that had traditionally been a feature of village water economies (Department of Health, 1995, p.764).

However, the 1962 cholera outbreak was not only a serious health matter, but also had major economic and political repercussions. Due to the outbreak, agricultural products such as bananas could no longer be exported from Taiwan to other countries, resulting in losses amounting to millions of US dollars, and exports of other agricultural products were also seriously hit at that time (Taiwan Provincial Council, 1962, p.343). The disease caused serious damage to Taiwan's national development both politically and economically since Taiwan was still heavily dependent on agricultural exports (Department of Health, 1995, p.384). Thus cholera combined with the strategic economic interests of a strongly export-oriented developmental state to accelerate the spread of safe water and sewerage systems in rural Taiwan.

Immediately following the cholera outbreak of July 17, 1962, four action group Cabinet meetings against the epidemic were held in August and all ministers of the Central Government attended to discuss the measures that should be taken to prevent the spread of the disease (Executive Yuan, 1962,a,b,c). Compulsory chlorination was extended to privately owned borewells as an emergency measure. A government report in 1964 revealed that only 31 percent of the population in Taiwan was then covered by tapwater systems and a further 20 percent of the population had access to safe drinking water through borewells and the 'simplified systems'. The remaining 49 percent of the overall population in Taiwan was regarded as being 'without access to sanitary water' (Taiwan Provincial Government, 1964). In the meantime, public tapwater construction that involved extending existing tapwater systems and building simplified water systems was carried out and incorporated under a new national economic plan. By 1970, 430 simplified water systems were built and at the time the public tapwater coverage increased to over 43 percent in 1970 (Department of Health, 1995, p.765; p1456).

The technological scaling up, in particular of rural systems, at the same time involved a rescaling of the governance of the emergent economies of water, including financial governance through the instituting of new price regimes along with the new water systems. In order to procure foreign loans and ensure that such funds were efficiently used, the Water Provision Foundation was set up in 1965 to plan the construction and management of tapwater systems nationwide. Exemplified by Taipei, access to foreign loans was made contingent on introducing new pricing systems of cost recovery. So, in 1961, the IDA representatives came to review the loan application in relation to the city's water system (Wang, 2001, p.1133). The representative proposed that a revised water rate plan be applied to the city as a condition for obtaining a loan contract from the IDA.

Consequently, Provincial Governments were made responsible for fixing and controlling the price of water through water rates. The rates for water had been unregulated by the Central Government, and each water system controlled by a local council had its own rules in setting water prices which differed according to different water rate systems. Most importantly, water had previously been treated almost as a free public good particularly in rural areas where people drew groundwater from their own borewells. In 1960, the Central Government set up a standard water rate formula that would apply to every water system. Although the water rates still differed according to the different running costs across local systems, it became a requirement on Provincial Governments to set water rates both to ensure cost recovery from the new systems, and to pay the interest on foreign loans.

Nonetheless, following the scaling up and standardisation of Taiwanese water economies, both urban tapwater and rural simplified systems, there was still a patchwork quilt of different scales before 1972. There were different types of public tapwater suppliers in terms of scale in Taiwan (see Table 6.2). Four inter-regional tapwater systems were set up and managed by Provincial Governments, while the smaller-scale county-based water systems, especially the rural simple systems, were run by the individual counties. The inter-local council water systems were water supply networks made up of various smaller systems across local cities or councils. Finally, there were also small council-based water systems which had been the legacy of the Japanese era. Table 6.2 shows the tapwater output based on the different types of water system in 1972. The table clearly reveals that the Provincial Government and the local county-based governments were the primary providers of water on the island in the early 1970s.

Table 6.2 Taiwan water systems in 1972.

Type	*Total N*	*Water output per day (m^3/per day)*	*Water output % (by column)*
Provincial run	7	347,000	36
County-based	8	477,000	50
Inter-local council	7	93,000	9.7
Local council-based	106	44,000	4.7
Total	128	961,000	100

Source: Wang et al., 2001, vol.5, p.1051
Note: Simplified systems are included in the local council-based category.

Viewed overall, the period 1945 to 1974 was one of the quite radical transformations of Taiwanese economies of water, especially the rural ones. New technologies of producing safe water, new scales of governance, new politically instituted processes of exchange through the standardising of a rating formula, were established. These economies of water were transformed by a matrix of forces, geopolitical, biological and environmental, and shaped by the internal dynamic of Taiwan's own economic development. The conditionality of international loans highlights the link between the geopolitical and the drive for extending safe water to the wider population. The continuing terror of cholera was conjoined with the developmental logic of economic growth of an export oriented economy. The political powers of a KMT dictatorship laid down the basis of standardisation in Provincial governance, led from the centre to secure water security and quality. Such was the complex and multi-dimensional character of post-war water transformations in Taiwan.

Creating a national public good: the unification of water systems since 1974

Following the developmental road of an export-oriented economy, rapid urbanisation and increasing demands for water then induced a further major reconfiguration of Taiwan's economies of water. From the 1970s, two major transformations, one related to technological infrastructures, the other to economic organisation and governance, combined to integrate Taiwanese water resources into a national public good. Still under KMT rule which witnessed a gradual transition from dictatorship to democracy only in the 1990s, the distinctive characteristic of this public good was state corporatist rather than democratic, driven by a rationality of central control, rather than one of democratic rights to a socialised common good.

In 1972, the Premier of the Executive Yuan of the Central Government argued that Taiwan's urban development required a single, centralised, water management system for the whole island, to spread tapwater from its then current level of 43 percent. In 1974, the Taiwan Water Corporation was established, and 128 local public water systems were incorporated between 1974 and 1976. Following the reorganisation of public water provision, the total number of public water supply systems grew quickly (see Table 6.1). The public water coverage was extended to over 90 percent of the total population by 1996 (Table 6.1).

At the same time, the source from which the water was obtained shifted dramatically due to increased water demand. River water replaced groundwater as the principal source for domestic and industrial provision. In 1977, 41 percent of drinking water came from groundwater and 36 percent came from reservoirs. In 1996, 26 percent came from groundwater sources and 60 percent came from reservoirs (Wang 2001, vol. 6: 816). Fourteen major water dams were built in the 1970s and 1980s to harness Taiwan's mountain tropical watersheds. The TWC became responsible for the water system of the whole of Taiwan excluding Taipei City which has its own water system. Coinciding with the government reorganisation in the 1990s, in 1999 TWC became a national state enterprise under the management of Ministry of the Economy, highlighting the linkage between economic development and a national economy of water.

Nonetheless, despite the achievement of near universal coverage of tapwater, providing good quality water continued to be a major challenge to water infrastructures: natural environmental shocks inflicted significant failures of supply over the last two decades. Typhoons and droughts, extreme environmental irregularities, resulted in serious water shortages every three to four years in the last two decades. In 2004, a typhoon with torrential rain resulted in the breakdown of public water treatment facilities in the northern region, Taoyuan, due to the turbidity of the untreated water. In 2005, the same event occurred in Taoyuan again and also in the Kaohsiung area (Taiwan Water Association, 2006b: 6–12). Both caused interruptions to water supply, and households were forced to obtain their water from other sources by pumping groundwater from borewells or buying water from water tankers. The limited storage facilities for treated drinkable water in Taoyuan could not even provide water for one day in the region (ibid: 6–16). In 2013, shortage of rain in March again caused the government alarm. In response to persistent irregularities of water supply over decades, each household in Taiwan has invested in either an individual or collective rooftop water tank. A striking feature of the Taiwanese urban landscape is the almost universal sealed rooftop water tank, exposed, of course, to the full heat of the tropical sun. Undoubtedly, this domestic infrastructural hedge against uncertainties of supply with its attendant risks to health, constitutes one of the underlying conditions for how the Taiwanese consume tapwater. Boiling water is an almost universal social practice, drinking water straight from the tap, almost unimaginable. Moreover, it is a social practice supported strongly by years of insistent government advice.

But, uncertainty of supply and the particular domestic infrastructures responding to it, combine to amplify the enduring risks to health from contaminated water. From the early 1950s and 1960s, public water had been highly contaminated by bacteria (see Table 6.3). In 1951, over 25 percent of public tapwater samples were found to contain *Escherichia coli (E. coli).* Only about 28.4 percent of tapwater quality met the water standard at that time. Although much improved by the

Table 6.3 Percentage of public water failing quality tests in Taiwan by region.

	Total	*Taipei city*	*Taipei county*	*Kaohsiung city*	*Penghu islands*	*Kinmen island*
1951	71.6					
1956	46.4					
1961	9.6					
1966	13.3					
1971	8.9					
1974	13.1					
1992	3.36	–	9.14	30.58	24	–
1995	4.28	7.14	0.92	6.14	42.55	–
2000	1.59	–	–	6.67	10.65	49.32
2005	0.25	–	–	–	0.64	19.27
2010	0.28	0.19	0.52	–	0.32	7.34
2011	0.28	–	0.77	–	0.25	9.60

Sources: Data for 1951–1971. Department of Health, 1995, Vol. 2, p.769; Data for 1974: Department of Health, 1995, Vol.5, p.1472, Data from 1991 onwards: http://ivy1.epa.gov.tw/drinkwater/index-10.htm

1970s, identified risks to over 10 percent of supply persisted, providing grounds for general lack of certainty in the safety of supply. These conditions doubly reinforced the necessity to boil water before consumption: necessity coinciding with the force of customary culture. In many households, domestic defences against risks to health and poor quality water are supplemented by water filtration equipment: water from the tap is often both filtered and boiled before being consumed. The number of domestic filtration units produced each year has grown from 7,322 in 1982 to 626,087 in 2012 (Ministry of Economy, 2013), and supermarkets display an extensive range of equipment, some with multiple high-tech filtration technologies – at a cost.

Alongside the market for domestic filtration, and equally shaped by Taiwanese context and culture, bottled drinks also became a significant commercial and market complementing both public tapwater and domestic filtration. Here too, Taiwan displays its quite distinctive characteristics, demonstrating the significance of the shaping and containment of markets by both competing alternatives and cultural norms. The emergence of dedicated bottled drinking water sold commercially as a market commodity began in the early 1980s. It was overwhelmingly 'packaged' water, mostly industrially processed water, abstracted from aquifers, rivers, tapwater and recently deep sea water. This commercial bottled water market is dominated by local beverage companies, especially by those with the national-wide retail and sales outlets. The market for commercial water expanded swiftly. In 1996 total sales of water reached NT$ 3100 million and doubled again by 2011 (Taiwan Water Corporation, 2011). A total of 364,087,000 litres were consumed in 2010, about 17 litres per person per year (compared to 36 and 146 litres in the UK and Germany respectively). The bottled water share was about 16 percent of the total consumption of non-alcoholic drinks. Sold in plastic (PET/PP) bottles in various quantities (0.6 litres, 1 litre, 2 litres, 5 litres, 10 litres), it meets people's needs for safe and technically healthy water, combined with convenience for mobility and immediate consumption. It is easily available from supermarkets and ubiquitous convenience outlets. Bottled water is just cheaper than soft drinks (NT$ 18 for 0.6 litres), but is far more expensive compared to other drinking water alternatives.

Yet, in Taiwan, in contrast to either Mexico City or Germany, up to now the consumption of commodity bottled water has been marginal, and its level of per capita consumption minimal. There is another mark of Taiwanese distinction. The Taiwanese drink two and a half times more bottled tea than bottled water, and bottled tea had been the major bottled drink product in Taiwan since the late 1980s. Bottled tea outstrips the sale of soft drinks by a similar margin, with soft drinks on a declining trend, bottled tea on a rising one. The consumption practice of drinking water as tea dominates the bottled market as much as it does the domestic consumption of water, boiled and filtered, out of a tap. And, as we shall see, bottled water in Taiwan has another distinctive competitor in the water garage market, buying filtered water from the pump.

Thus, the bottled water market in Taiwan finds a distinctively minor place in the development of economies of water in Taiwan. The consumption work of boiling and filtering typical of Taiwan, together with cultures of tea consumption for drinking water, are thus now plumbed-in to domestic and public infrastructures, conditioned by tropical environmental conditions, including the watersheds from Taiwan's central mountain spine. It is beyond assessment to gauge the relative significance of the fear of cholera, the risks of bacterial contamination, the uncertainties of supply, or the distaste for chlorine. Societal memories mix and meld different anxieties and fears. Moreover, a dominant cultural aesthetic for boiled water and tea cannot be reduced to a response to such fears, counterposing the positive aesthetics of taste and preference to the negative reaction to biological threats. Tea is not drunk *because* it is safer than raw tapwater (although in Taiwan it has been): historical and cultural fusion between preference and safety does not allow for weighing the one against the other in the social practice of drinking tea, whether produced domestically or purchased in bottles.

The 'spinal core' economy of water, now supporting the majority of the population of Taiwan is summarised in the configurational diagram below (Figure 6.1), distinguishing the island in terms of its economic organisation of production, distribution, exchange and consumption. Taiwan had developed a distinctive and integrated state-corporatist production system, a unified price-exchange institution, characteristic domestic infrastructures, entrenched practices of boiling

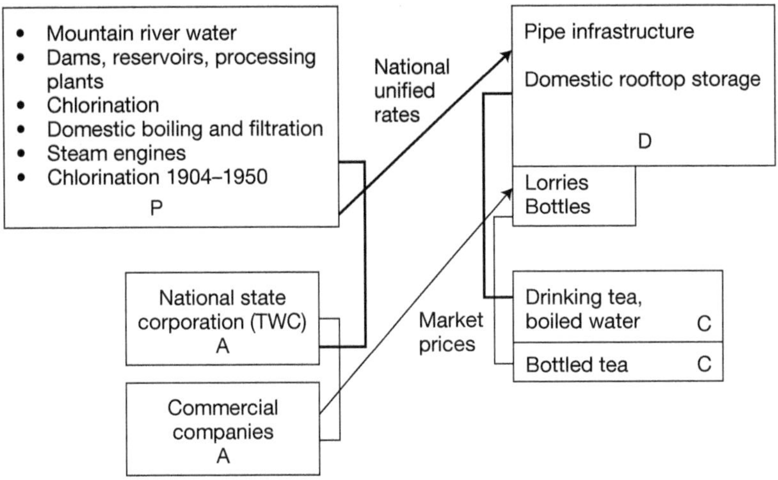

Figure 6.1 The spinal core economy of drinking water in Taiwan.

water, and a dominant mode and culture of consuming water as tea. Yet, as we have seen, this current configuration is only a moment in a long and complex process of historical transformation.

Kaohsiung and Pingtung

Zooming in to a region of Taiwan, the city of Kaohsiung and Pingtung County, two significant variant economies of water have emerged, again distinctively Taiwanese, as a consequence of uneven development, and more extreme risks and uncertainties attached to tapwater. Kaohsiung city is located in the south part of Taiwan as one of the major industrial cities since the 1960s with a total population over 1.5 million at the beginning of this century. In 2010, the city merged with the surrounding county becoming Kaohsiung Metropolitan City with a total population over 2.7 million. The public water system now covers over 90 percent of the total city population. Pingtung County, with a mostly rural population of 855,000, has a tapwater coverage of only 45 percent, mostly concentrated in urban areas, including Pingtung City.

The Kaohsiung variant economy of water

If weak trust in the reliability, safety and quality of tapwater on the island of Taiwan underpins practices of domestic filtering and boiling of water, Kaohsiung stood out with its much higher levels of contamination. In 1992, 30 percent of water samples from Kaohsiung City failed the test compared with only 3.36 of the total of the Taiwan area water sample that did not meet the quality standard at that time (Table 6.3). In 2001, TWC water samples from Kaohsiung failed 40 times and the corporation was fined NT$ 60,000 each time. Even so, the following year they failed to pass the requirements 19 times (TWC, 2013). In these circumstances, a dual economy of water emerged; the tapwater system primarily for non-drinking uses, and a separate, market economy for dedicated drinking water, serviced by water stations or 'garages'. Both economies have responded to the crisis in quality and trust of tapwater, and have emerged in complementarity. In 2010, a total of 197,399,019 tons of tapwater was provided by the TWC, with an average 357 litres per person per day (WRA, 2010: A-14). On the other hand, the city's drinking water has been mainly provided by the water stations for over a decade. It has been estimated that 80 percent of the total population bought their drinking water from water stations in 1997 and each person consumed two litres of drinking water per day (Wu, 1997).

In response to the quality crises, the public infrastructure for tapwater under the TWC has undergone major modernisation in the last decade. Water had been drawn upstream from the main river flowing through the city, and, up to ten years ago, there had been a major problem of pollution from riverside pig farms. The government had for many years subsidised piggery sewage treatment, but this proved inadequate, especially after subsidies were reduced. Consequently, ten years ago, in a dramatic intervention, the piggeries and six million pigs were compulsorily removed from the risk region. At the same time, from 2002, there was a major upgrade of the whole treatment system, owing to consumer dissatisfaction at the high levels of limescale, smell of chlorine and hardness of the water, which made the water turn milky when boiled. A modern water treatment plant has been built, with multi-stage purification technologies, involving ozone treatment, biologically activated carbon filters, and water softeners. Before 2000, chlorine was added only once after the water had been treated in order to protect against e-coli contamination in the pipe network. High levels of chlorine remained in the water, resulting in a strong chlorine smell of the water from the tap. The new system has introduced chlorine checkpoints in the network, reducing chlorine levels at the endpoint for domestic consumption.

Nevertheless, given the recent history of contamination and poor water quality, Kaohsiung's general public considers that TWC water is not fit for drinking. From the 1980s, private drinking water economies have flourished in this modern city. Private run water 'stations' sprung up all over the city. There are now officially registered 1,771 such stations, over 1,000 of which are located in the central urban area (Kaohsiung Department of Health, 2013).

From the 1980s, the early form of the water station was sourced from mountain springs, transported by water tankers into the city, supplying water pumps in quite small-scale water garages. The water source, being national state property, was extracted under license. A cartel of eight deep wells each with a private owner, is located in Pingtung County, producing over 4,717,260 tons of water per year (Water Resource Agency, 2013), enough to support a population of 6.4 million a year if each person takes two litres of water a day. They provided one of the main groundwater sources for Kaohsiung's water stations, and the cartel fixes the price, which, at the time of interview in March 2013, was $20 a ton. So, publicly owned groundwater is exchanged under license to private deep well owners, and then sold on to consumers. Some of the tankers and the stations are directly owned by the well owners, while in other cases, the water supply chain is less integrated. In terms of appropriation, a succession of transfers of ownership, from public to private, and then from private to end consumer, characterises this early development of the water station economies in Kaohsiung. Moreover, it is a state-regulated end market, with water quality regularly inspected and certified by the local health department. However, given the consumption work of collecting the water from the stations in 20-litre plastic containers, and then storing it in the home, local authority advice is for all water from water stations to be boiled before being consumed. Consumers thus play an active role in the distribution system, both collecting water often on an almost daily basis, and treating water prior to consumption.

From the early 1990s, this spring water station business in the city has faced competition from water stations using TWC tapwater as the water source. Directly connected to the mains water supply, the water station owners pay $9–11 per ton for TWC tapwater, the same as charged for domestic households. These new water stations are more organised and standardised in their water management. The technologies of filtration are on display, often with shiny stainless steel piping, flashing blue lighting, and ranks of different filters each with a 'scientific' label and purpose: UV and RO (ultraviolet and reverse osmosis), Ph for hardness and softness, and some of the pumps offering mineralised water. Advertisements for the water stress how technically and scientifically pure and healthy the water is; a complete antithesis of equating 'healthy' with 'natural'. In terms of processes of 'qualification' of water, therefore, a set of water pumps offers different qualities of water, from different batteries of filters, for different prices, from RO only, through to mineralised and carbonated water with the full battery of filters. Interviewees told us that different water qualities were chosen for different purposes, the middle-ranged price water being popular among tea drinkers, tea shops and restaurants. The prices paid for this high-tech water are up to a staggering 333 times higher than the cost of its original tapwater source, although RO technologies involve high levels of water wastage, yielding as little as 1 litre from every 5 litres of tapwater. Even if one takes 1 litre of RO water as equivalent to 5 litres of tapwater, the price of the most expensive garage water is 660 percent higher than tapwater, and the cheapest over 200 percent higher (see Table 6.4).

The branding of these new TWC water stations emphasises the filtering and quality-enhancing processes improving tapwater, in contrast to the early water station forms that were essentially a system of distribution of spring water into the urban areas. The modern form is expanding more rapidly, and also scaling up, with firms owning chains of outlets, not only within Kaohsiung, but in other urban areas in the South West as well. Several of the water station companies are also vertically integrating into filtration equipment manufacturers, assembling the display cabinets for their high-tech water products.

Kaohsiung has therefore witnessed the emergence and instituting of a variant economy of dedicated drinking water, the water garage system (see Figure 6.2). Developed in response to lack of trust in tapwater, this economy of drinking

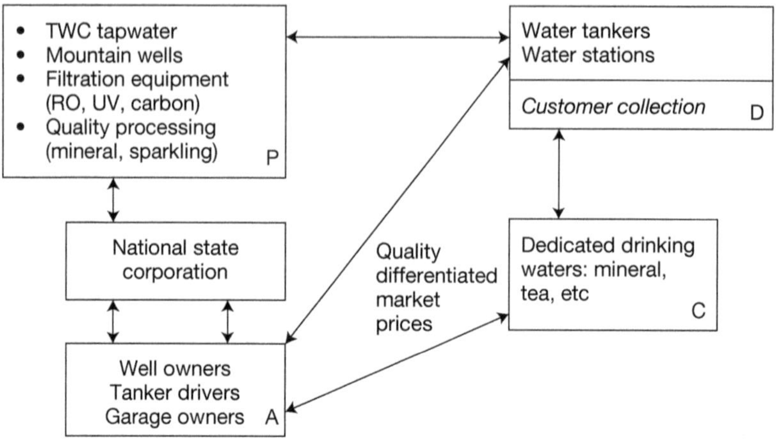

Figure 6.2 The water-garage economy of drinking water: Kaohsiung City.

water has formed markets and distribution systems peculiar to Taiwan, and involved exchanges between national or state-corporation public water authorities and market actors before being sold to end consumers. A quite distinctive, almost anti-natural, process of quality-of-water differentiation forms an integral aspect of this economy, where premium prices are paid for high-tech 'synthetic' water, drawn from an 'ordinary' tapwater source. Linked to the distinctive mode of distribution from water pumps at water garages, however, there is a persistent quality paradox: however technically pure the water is when distributed and sold at the pump, customer collection and storage in plastic containers entails a risk of contamination, so still requiring the ubiquitous consumption work of boiling before use. Conversely, as a particular economy of drinking water, it provides a markedly cheaper alternative to bottled water (see Table 6.4), in a possible trade-off between cost and safety. Even in Taiwan, people do not generally boil bottled water – although they do prefer bottled tea.

The rural Pingtung County variant water economy

Pingtung County is adjacent to Kaohsiung city, and has a rural economy with 25 percent of households working in agricultural production. With tapwater available to only 45 percent of the population, mostly in urban areas, a water tanker delivery economy has emerged as the dominant source of water for drinking and cooking. Moreover, the traditional countryside source of private borewells has been damaged by fertiliser and pesticide pollution from agriculture, so that, at most, groundwater is used for washing and cleaning. A typical case of agricultural economy–nature interactions impacting on natural resources, degrading them for human use.

Although water garages have been opened in many villages, home delivery of water by water tankers, distributing the water from the same deep well mountain sources that supply Kaohsiung's water garages, has become the dominant domestic economy of drinking and cooking water. Each household has an external, ground-level cistern, routinely re-supplied by weekly or twice-weekly delivery rounds. The intermediaries between the well owners and the consumer are self-employed tanker-owner drivers. Competition with tapwater water stations in Kaohsiung has driven down the price of water from the well owners to $10 to $15 per ton, resold to villagers by the tanker drivers at $10 for 20 litres. One water tanker can cover five or six villages in one week if they go to one village once a week. From fieldwork interviews, typically two tanker drivers compete for the water business in a village of 400 households. Thus, in this rural water economy, the power asymmetries of exchange in the supply chain favour the well owners, with tanker owners and consumers both being price-takers rather than price-makers. Consumers pay $2.50 per litre for drinking and cooking water, as against the $0.012 paid for tapwater in Kaohsiung city, and the rest of Taiwan (Table 6.3). However, given the domestic appropriation of groundwater for washing and cleaning, as a 'free' private good, there remains a widespread mindset of resistance to the installation of piped tapwater infrastructures, where water has to be paid for, even at a low price, when it too is deemed non-potable.

Thus, Pingtung County has developed its own variant economy of water, at the margins of both the national system and the Kaohsiung alternative garage economy. Drawing on the same national sources that also supply Kaohsiung's stations, it is marginal economy dependent on, yet also reacting against, the dominant Taiwanese configuration.

Table 6.4 Costs of drinking water in different Taiwan water economies.

	Purchase from source per litre	*Price to consumer per litre*	*Price to consumer/ price to trader*
Public tapwater (TWC)		0.009-0.012	
UFO water station	$0.012 from TWC	$4.0	333x / 5 = 660%
RO, UV, carbonated		$2.0	166x
RO, carbonated		$1.5	125x
RO only		$1.33	110x / 5 = 202%
Kaohsiung water station	$0.02 from 'spring'	$1	
Pingtung tankers	$0.02 from 'spring'	$2.5	125x / 5 = 205%
Cheap bottled water		$14	

Source: interviews in March 2013 with well owners, tanker drivers, station owners, and fieldwork observations

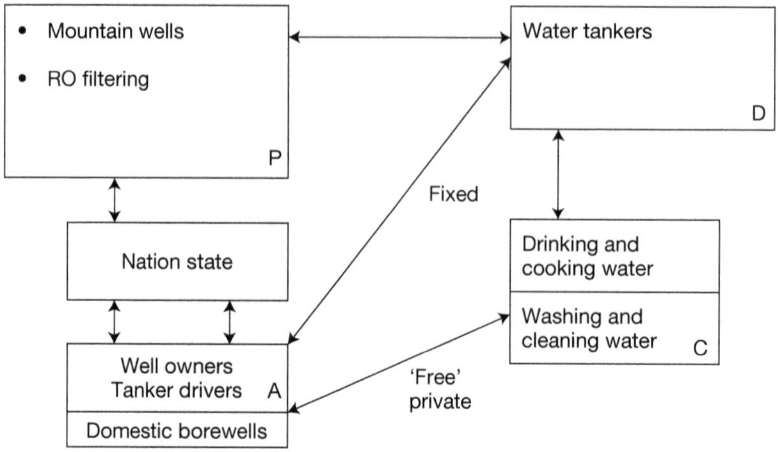

Figure 6.3 Pingtung County rural economy of domestic water.

Taiwanese trajectories

Taiwan has seen the emergence of a 'spinal' national economy of domestic water, shaped by Japanese occupation and KMT dictatorship for much of the twentieth century, and inflected by its geopolitical significance in the Cold War. Taiwan's economies of water thus make conspicuous the significance of political structures in economies of water, more subterranean and covered by layers of history in other national societal contexts. At the same time, the politically oriented instituting of economies of water in Taiwan was conditioned by environmental and biological resources and risks peculiar to a sub-tropical island. Environmental and biological shocks (typhoons and cholera) triggered major shifts in central state direction in developing economies of water. Even now, the sustainability of the central spine economy of water is at risk on the one hand from the typhoon-induced silting of dams and, on the other, popular resistance to more dam building (Peng, 2013).

Uneven development of the spinal water economy, especially in the South West, created the context for further, Taiwan-distinctive economies of drinking and cooking water, in Kaohsiung city and Pingtung County. Whether sourced from 'nationalised' mountain spring water, or from the state corporatist Taiwan Water Corporation, these subsidiary economies of water were interdependent with and contextually shaped by, the spinal economy of water. They emerged over several decades either in the context of low trust and inadequacy of public tapwater or in the absence of tapwater infrastructures. The water predilections of Kaohsiung consumers for high-tech synthetic water exemplify a quite general rejection of the standard equation, especially for Western bottled water markets, of 'natural = good, healthy, pure'. Nature is not given as 'safe' in a sub-tropical context, and industrialised tapwater systems linked to domestic plumbing infrastructures are less hygienically secure. Consequently, the final phase of water 'qualification' across all the economies of water in Taiwan is to boil it, and flavour it. This fundamentally conditions how the Taiwanese drink water, whether 'neat', or in tea and soups.

The different configurations of production, appropriation, distribution and consumption characteristic of contemporary Taiwan thus bear the mark of their historical formation. Distinctive forms of appropriation of water, by a centralised nation state and its state corporatist organisation, the Taiwan Water Corporation, are combined with distinctive forms of distribution – notably, the water garages of Kaohsiung. Different combinations of property rights, market and non-market emerge and interdepend with each other, in the context of particular forms of state power. Patterns of water consumption are linked to dual systems of provision of different waters for different purposes: cooking, drinking, washing and cleaning. Taiwanese consumption of water is relational within these configurations, which, in turn, are embedded in natural environments which leave deep resonances of fear in societal memories of cholera, droughts or deluges. Practices of water consumption are conditioned by both societal *and* natural contexts, where risk is as much natural as economic or political. Indeed, practices are formed under the pressure of fusion of these dimensions, arising from the particular and successive society–nature interactions in the historical provision of water from the Japanese epoch through to the present.

7 Water variations

As a substance, water to drink appears to be both basic and simple; and, as a discrete act, drinking water out of a glass likewise. Yet, whether across Delhi and Mexico, the United Kingdom and Taiwan, Germany and Italy, there is no singular common substance, and whatever it may be today has been the product of long and complicated histories, none of which have arrived at a final terminus. Likewise, how people drink water, in what form and in what social routine or practice, has varied, is varying, and no doubt will vary, in time and space.

The comparisons undertaken in this book are selective. First of all, the focus has been on urban drinking water, largely to the exclusion of rural populations. Secondly, with the partial exception of the analysis of Taiwan and Mexico City, major issues of competition for water resources between domestic, agricultural and industrial uses – highly significant in other society-nature contexts such as Australia – have been ignored. Moreover, the selection of cities and countries is to an extent inevitably arbitrary and opportunistic. However, the reason for using a comparative and historical methodology here has the overriding objective of stretching analytical concepts, rather than providing a set of cases, then to be placed in a taxonomic grid, as is common with many comparative approaches (Harvey, 2014b). We have been looking at processes of open-ended change, how water has been, and how it is becoming, with no assumptions of arriving at any settled water end-points. The analytical focus is on processes of variation, rather than any given historical or contemporary outcomes of variation.

As stated in the introduction, drinking water is both an object to be analysed in its own right, and a probe to explore major issues of political economy, the role of markets, collectivities, and the state; sustainability crises and economy-society-nature interactions; and the dynamics of transition and structural change. Unsurprisingly, all writers on water have stressed the basic and universal human need for potable water. However, even the Millennium Development Goal does not prescribe either what counts as access or what constitutes 'potable', and historically and across our cases, there are varied meanings and varied outcomes of what makes water potable (Gleick, 2004, 2011). In the municipal hydraulic model, it has only very latterly been seen as a public health good, chlorinated, and guaranteed 'safe' at the mouth of every household tap – almost as if that were the only quality that mattered (Parker and Sewell, 1988). So, as a non-discretionary good, something that humans cannot biologically survive without whatever form it takes, water as a probe asks many pressing questions of the relation between societies and nature, of the character of polities, and of economic organisation including cultures of consumption. Societal modes of sustaining water provisioning have to be found for continued social existence – as was only too clearly illustrated by the collapse of Mayan civilisation. Societal variations are deeply connected to water variations, and vice versa.

As a probe, different societal and historical trajectories of drinking water ask testing questions of the nature of both commodities and public goods – as Bakker (2003a) has noted about water as a characteristically uncooperative commodity, to which we have added water as a problematic public good. In attempting to answer these questions, we need to see that forms of economic good evolve, hybridise, and innovate into new forms of economic good, transcending static classifications and dichotomies. For example, from our analysis it is clear that drinking water does not become a commodity just by virtue of people having to pay for it, even when the provider is a multi-national corporation. And yet, there are some Irish organisations that consider drinking water from the tap is only a public good when paid for out of general taxation. Moreover, there is, in any economy of water, a cost in terms of labour and time in organising sufficient and consistent supplies of water fit for consumption (however it might be consumed), often involving divisions of labour and exchanges, even when these are not monetised. Water, as a probe comparing its varied societal change across time, stretches concepts of economic good.

Cutting across the evolving character of water as an economic good, the conceptualisation of processes of qualification is also stretched by considering the contrasts between bottled and tapwater. Is the process of qualification of tapwater, and indeed the quality regulation of tapwater, instituted or normalised in the same way as bottled water, and how do processes of qualification vary across societies, and historically? Adopting a position that qualification needs to be understood holistically, 'from source to sip', immediately challenges the dominant discourse which stigmatises bottled water as if it was more comparable with tapwater than apples are with pears (Barlow and Clarke, 2002; Gleick, 2010; Salzman, 2014). To illustrate the force of this point, we have not encountered people making stilton and broccoli soup or cups of instant coffee with sparkling bottled water, whereas we have with tapwater. Not only the quality of the substance but also the way people prepare and consume it differs, and those differences are systemic when exploring their societal variations. The quality ascription of 'safe' or 'drinking water', when applied to bottled or tapwater masks different processes of qualification. The assumptions behind some questions contesting why people buy bottled water when they can get tapwater virtually free, or when they cannot taste the difference between some bottled waters and some tapwaters, are deeply flawed by treating them as if they were essentially the same thing: safe and aesthetically fairly neutral.

A strand running through the comparator cases are the questions water poses concerning the relation between economy and polity. Is there a connection between the emergence of universal suffrage and a universal right of access to, and equity for, drinking water? Disempowerment and water deprivation in Delhi or municipalisation in London in the early twentieth century speak to this question. But, military occupation and dictatorship in Taiwan point to quite different underpinnings of the universalisation and 'nationalisation' of water supplies.

Finally, all of our cases highlight the significance of particular polity-economy-nature interactions in the trajectories of development of drinking water provisioning in the variety of water-environmental contexts from sub-tropical Taiwan to the temperate zone island of the UK, and from Mexico City without a river to Delhi with a contested and contaminated one. Questions of economic growth and urbanisation in these different environmental contexts display different sustainability dynamics, different sociogenic sustainability crises. So drinking water, as a comparative probe, is good for thinking about the significance of resource sustainability in the dynamics of changing economies of water.

In this final chapter, we consider the theoretical implications of historical and comparative analysis of drinking water in terms of three broad themes: transitions and reconfigurations; processes of qualification; and society-economy-nature interactions, sustainability and sociogenic crises.

Transitions and reconfigurations

The IEP approach enters into the analysis of transitions, or historical transformations, through the lens of economies of water without, however, a priori privileging economic dynamics over cultural, political, other social, or even natural-environmental dynamics. It is a choice of a point of entry into a complex of interactions, involving multiple dynamics. That said, the anthropological take on what constitutes economies – relational configurations of production, exchange/appropriation, distribution and consumption – entails a notion of transition as configurational change, including, but extending way beyond a technological change such as the emergence of a piped infrastructure for water or for sewage. Across all our water variations, transformations of economic configurations have occurred in co-evolution with transformations of political structures, and in the scale and nature of exploitation of natural resources, ground and surface water. So although initially choosing economic re-configuration as the significant mark of transition, our analysis of each of our cases has taken the overall economy-polity-nature relational matrix as the object of analysis for systemic historical variation.

Transition theory has provided some fruitful analysis of the complexity of socio-technological transitions, insisting on a multi-level approach, from small-niche, bottom-up, innovation, through socio-technical regimes involving multiple actors and inter-dependent technologies to the overall landscape of the social economy (Geels, 2002, 2004, 2006; Geels and Schot, 2007). Some comparative analysis has been undertaken, but, in general between singular transition cases, and never in the context of societal variation. Comparing the transitions from horse-drawn vehicles to automobiles, sail to steamships, or piece work to continuous flow mass production, or cesspits to sewage systems, four different types of pathway were identified (Geels and Schott, 2007). Thus, a de-alignment/re-alignment pathway was ascribed to the transition from horse-drawn carriages to automobiles in the USA, in which major changes in the societal landscape (urbanisation, immigration, electrification and expanding middle class with disposable income) created the conditions for competition for alternative modes of transport for short and long distances, eventually leading to the dominance of the automobile. The other pathways are distinguished primarily on criteria of the timing and direction of disruptive pressures, whether from the landscape, regime, or niche level. Thus, the technology substitution transition pathway was exemplified by the sail to steamship transition, where a sudden, landscape-induced increase in trans-Atlantic travel for mass migration created an opening for an already developed steamship technology to rapidly take over the sea transport market. This comparative typology certainly increases the scope and flexibility of the multi-level perspective model. But the limitations of this 'singular case' approach to transition in a singular societal context, are manifest in the example closest to the concerns of this book, the transition from cesspits to piped sewage infrastructures in the Netherlands. It is typologised as belonging to a 'transformation pathway', where external pressures from hygienic movements and changes in cultural expectations and norms of cleanliness induced the abandonment of the technological cesspit regime in favour of sewage systems, but only in the 1890s in The Hague and 1914 in Amsterdam (Geels, 2006). This was quite remarkably late in comparison to London, and even, as discussed in Chapter 1, to Hamburg. As analysed in Chapter 2, very different transition processes to technologically similar outcomes occurred within the UK, and they were only a staging post to further major transitions, as a consequence of a major sustainability crisis which was both biological and political. By the mid-nineteenth century, given available know-how and the surface similarities in technologies of sewage systems (as against contrasts in political or economic organisation), the significance of societal and specific nature-polity interactions must enter into the explanation of the huge temporal spread of when these transitions occurred in different societies. It is a question pressingly posed by Delhi today, when analysing the dynamics of a blocked or impeded transition.

Moreover, from its provenance in science and technology studies, transition theory's main focus has been on transformations of socio-technological regimes, less on the economic dimension of transitions (forms of socialisation or privatisation, for example), and less still on political dimensions (changes in forms of rule from dictatorships to forms of representative democracy, or vice versa). Evans' analysis contrasting Hamburg with Prussia (Evans, 1987), or transitions of political institutions within London (Trentmann and Taylor, 2005; Taylor and Trentmann, 2011; Broich, 2013), further amplify the significance of these dimensions arising from comparative and historical societal analysis.

The empirical material assembled in this book bearing on the analysis of transitions immediately suggests a broad contrast between what might be termed endogenous and exogenous transitions, and, for the analysis of endogenous transitions, a long duration approach to successive transitions, particularly in the case of London. The economies of water in Taiwan were transformed by two successive incursions: Japanese invasion, and the military KMT dictatorship from mainland China, creating a unique geopolitical context for the subsequent trajectory of development. Mexico City is marked to this day by the effects of the original Spanish colonial war and colonisation of the sixteenth century. Delhi experienced first its absorption into the Moghul and then British empires, both also leaving legacies conditioning subsequent evolution. We will first seek to draw some analytical conclusions from the long duration endogenous succession of transitions, before turning to exogenously induced transitions, and their sharply contrasting dynamics of change.

Over the course of 400 years, from when the New River Company first laid down wooden pipes carrying water for 40 miles in 1614, five major reconfigurations of London's economies of drinking water were identified. From that arbitrary historical starting point, configurational analysis revealed the interdependence between changes in production of water (the switch to upstream river sources), appropriation and exchange (rights over extraction, a new system of consumers paying by property rateable values), distribution (the wooden pipes, domestic plumbing) and consumption (e.g. the gradual spread of tea drinking, more domestic washing). Royal charters and parliamentary statutes were also innovating new political regulations and institutions surrounding water. Finally, the beginnings of the shift from groundwater to surface water, with the growth of London's population, signalled a changing relation between new economies of water and environmental resources. Each of the successive reconfigurations displayed their own distinctive recombinations of production (new technologies of processing and qualification of water, activated sludge treatment), distribution networks (such as the link-up between sewage and piped water mediated by water closets), appropriation (e.g. new forms of control over resources under regulated private monopolies, evolution of water rates to address the emergence of the domestic bath), distribution (including innovations in domestic plumbing), and consumption (changing norms of bodily cleanliness, drinking water straight from the tap, instant coffee and teabags, etc.). Co-evolving with these economic reconfigurations, however, was the emergence – political transitions – of new forms of governmental institutions, intimately related both to innovating new forms of financing large-scale infrastructures and regulation of the new forms of water-economic activity and agents (e.g. pricing and quality). And implicated in this matrix of transition, each phase also marked a shifting place of political economies of water in their natural environments – symbolised in one of the transitions by Bazalgette's six dedicated barges dumping London's untreated sewage into the North Sea for nearly a century (both a physical and a political process stemming from the regulation of the day).

Comparing successive transitions, it is striking both that each transition is a reconfiguration of a previous transition, and that each transition is characterised by very different dynamics from the one that preceded and succeeded it. Thus, in the broad sweep, if we take the transition from the mid-nineteenth century private statutory water monopoly provision and public sewage to London's late municipalisation of water and sewage, it was argued that the dynamics were a combination of inequitable price differentials between companies and boroughs; uneven reliability and quality of supply (especially in drought years) between adjacent boroughs; social and political movements; all in the context of a major reform of the city's governance. The subsequent extended transition from municipalisation to privatisation, and the return, if in very different economic form, of private monopoly provision, was analysed in terms of the new dynamics arising from a number of interlinking configurational tensions: the disjuncture between integrated river basin management and city-based governmental institutions (LCC or Metropolitan Water Board); the de-coupling of financial and water-sewage flows, as water rates proved insufficient to meet the needs of substantial investment in infrastructure renewal to meet new EU environmental standards; and a fiscal crisis of central government over the Public Sector Borrowing Requirement. These destabilising dynamics, of course, coincided with a general political shift towards privatisation of previously socialised industries, but, again the political instituted 'markets' bore the legacy of prior socialisations in their organisation. The most dramatic example of the internal dynamics of a configuration triggering its own demise was illustrated by the transition to cast-iron pipe water infrastructures along with the spread of water closets and a law requiring their discharge via the sewers into the River Thames: the consequent Great Stink provoked the next major reconfiguration in sewage infrastructures *and* the capital's political organisation. Taking these three cases together, therefore, a significant part of any explanation for any given transition involves the dynamics, economic, political, and environmental, instituted by the previous transition.

A major theoretical consequence of this analysis is to move away from any general descriptive device, and to root explanation of transitions in their historical and spatial contexts. To some, this may appear to be a move away from abstract theory, or general models of explanation. But, of course, the argument for historically contingent and spatially contextual explanation is a highly abstract and theoretical one, as set out above. It is a view of immanent and historically emergent causalities, rather than general laws, or even laws pertaining to a whole historical epoch, such as industrial capitalism.

This leads us directly into one of the most interesting dynamics of the successive transition series: the processes of scaling, economic, political and environmental. Viewed in hindsight, the trajectory of London's water provisioning seemed set teleologically on a path to ever-expanding scale. In the first period, London's water economies were extremely localistic, wells, cesspits and piped water networks serving only districts or subdistricts, with no integration or coordination between them, whether economic or political. Thus, in the early nineteenth century, even the water wars between emergent water companies involved competition only on particular streets or squares, no London-wide competitive market or economic organisation for piped water supply. In the final period, we find a London internally integrated, but within the much enlarged scale of the Thames River basin, with political, environmental and economic regulation and coordination embracing many towns and cities aside from the capital itself. The politically defined scale of the River Water Authorities formed the basis of the subsequent scale and structure of the privatised water monopolies.

However, the teleological view is highly misleading. One of the significant elements of the dynamics of successive transitions was what could be described as scale-friction between economic organisation, political organisation, and environmental geography. In the early nineteenth century, for example, the political organisation of the vestries was often in conflict with the Commissioner of Sewers responsible for flows across vestry boundaries. Likewise, the mergers and acquisitions of water companies, swapping districts between them, established economic scales exceeding local political authority, so stimulating the coordination of vestries in the Anti-Water Monopoly Association. The up-scaling of the early nineteenth century through the extension of piped water, sewage, and WC networks, then led to further scale conflicts. It was not just that shit travels along the Thames and rivers flowing into it, across political boundaries, ending up under the nose of the national parliament. Significantly, the subsequent construction of the sewage intercept infrastructure by Bazalgette not only spatially connected and integrated London over a vastly greater spatial scale, but also involved new instruments of finance, and the instituting of new political structures, the Metropolitan Board of Works. Yet this too cannot be seen as straightforward scale harmonisation between political, economic and environmental scales. It led to a contrasting, and eventually conflicting, friction between the scale of operation of the sewage system and of the divided city of private monopoly water provision, only then superceded by municipalisation and integration under the Metropolitan Water Board. Finally, the emergence of River Water Authorities in the 1970s, and the 'nationalisation' of water resources in turn created friction between the scale of water and sewage provision management and the political and fiscal scale of local and metropolitan governments, a scale disjuncture between water and fiscal flows.

It could be argued that the contemporary instituted scales of the privatised water monopolies will prove a significant obstacle to state interventions to develop water transfers from the water-rich population-scarce North West to the water-poor population-dense areas in the South and East. Moreover, a direct consequence of this scaling has been the abandonment of the formation of a national scale for pricing water, common to all UK citizens, with cross-subsidisation. Prices paid for water in different regions have diverged even more under privatised monopolies. In scale conflict terms, one has only to contrast the scales frozen by the instituted privatisation monopolies within the tiny frame of the United Kingdom, with the scale conflicts involved between the central state in China and the source and transit provinces along the route of the South-North Water Transfer Project, in order to provide water for Beijing. In the United Kingdom, it is an open question whether price inequities, climate change and population concentrations might destabilise the current river basin scales, frozen by privatisation.

Reflecting on the scale dynamics of the successive transitions, one of the striking implications is that there is no pre-established multi-level scale, as in the Multi-Level Perspective, but rather the transition processes themselves were involved in constituting differentiated scale levels, economic, political and environmental. Moreover, there is nothing to suggest that scale-level differentiation is neatly divided into three (niche, regime, landscape), or that any scale-level differentiation is historically enduring.

So, this comparison between successive and serial, endogenous transitions has thrown light on the nature of the causal dynamics and explanatory frames adequate to them: historical, spatially contextual and scale formative. Turning now to instances of exogenously induced transitions, the cases of Mexico City, Delhi and Taiwan involve many of the endogenous dynamics just discussed, but in addition some very different dynamics which have, at key historical moments, changed the course and conditioned subsequent trajectories of political economies of water.

All three instances of exogenous transitions resulted from colonial and military occupation by a foreign power, yet they are nonetheless sharply contrasting cases. In Mexico City, the impact of the initial war of colonisation resulted in the draining of the mountain basin in which the contemporary city is situated. This 'one-off' act of destruction of the sophisticated Aztec regime of managing water resources nonetheless conditioned all subsequent endogenous development of the city, its water provisioning, its sewage discharges, and its frequent and widespread land subsidence and periodic flooding. The reliance on pumping water from the Cutzamala and Lerma river basins to a height of 2000 meters, with its consequent environmental impacts and energy consumption, including heavy reliance on 'importing' bottled water (Delgado-Ramos, 2014), is an outcome of several successive endogenous transitions all conditioned by an external intervention occurring 500 years ago.

By contrast, Delhi's historical water trajectory has been marked by two imperial invasions, Moghul and British, each of which introduced or transferred infrastructure technologies radically novel to the city at the time. The complex Moghul gravity-fed systems of canals, wells, vast storage cisterns, and cultures of social activities around baoli (stepwells), still leave their traces on the contemporary city. But, these infrastructures had already decayed, even collapsed, through internal decline and warfare with the British by the mid-nineteenth century. The British then imported the cast-iron pipe networks of pressure-distributed water of nineteenth-century Europe, entirely alien to the then Indian context, to supply the ruling elite and the upper strata of Indian society. The British colonial regime established the sharply divided economies of water, where public goods water to this day remains predominantly the privilege of the middle and upper strata. Even those good citizens who pay their below-cost water bills, legitimately purchasing water supplied by the Delhi Jal Board, exercise a privileged right to a public good unavailable, indeed in many cases prohibited, to millions of the poorer citizens of the capital. This remains a defining characteristic of 'public goods' water in the city, with its lineage stretching back to pre-Independence India (Prashad, 2001; Hosagrahar, 2010).

Taiwan provides the sharpest contrast to endogenous transitions, with first the military occupation by Japan earlier in the twentieth century, and then the military dictatorship of the Chiang Kai-Shek and the KMT stretching into the 1990s. The introduction of the hard technologies of piped water and the softer technologies of hygiene police and regulation under the Japanese was combined with the 'nationalisation' of Taiwan's water resources, including the genocidal appropriation of mountain sources from ethnic populations. Many traces of the Japanese infrastructure remain to this day, some still operational. This already strongly authoritarian and centralised restructuring was subsequently developed under the KMT dictatorship by a form of state corporatism (the Taiwan Water Corporation), with a nationally integrated economy of water, nationally uniform prices, and national quality regulation and monitoring. Additionally, much of the investment in developing the water infrastructures was provided by the USA, as a strategic element of their support for Taiwan in the geopolitics of the region, especially in relation to the People's Republic of China. Even if there remain pockets of relative underdevelopment, the transitions and trajectory of Taiwan's 'spinal core' of water provisioning illustrate the tight connection between structures of power and instituted economies of water, including the distinctively statist characteristic of 'public goods' water. Taiwan's successive transitions were not only induced exogenously, but also by structures of power associated with invasion and occupation. This contrasting case, however, puts into relief the significance of the very different English evolution of the political system of parliamentary democracy, and the slow emergence of universal suffrage, that ran parallel to the successive transitions of London's economies of water.

State and market, public good and commodity

The historical and spatial comparison of different trajectories of transition in economies of water leads to a very different approach to the question that has dominated much recent discussion of the relation between water and capitalism. What kind of economic good is tapwater, or, for that matter, bottled water? In this comparative work, we are only examining the question within a variety of broadly capitalist political economies, not self-styled socialist economies or future utopian ones. This is a necessary point to make, not only because state-owned enterprises in capitalist economies are often legally required to operate on a commercial basis, but also because, even when water provisioning is undertaken by a state-owned organisation and, for example, financed out of general taxation as in the Republic of Ireland, the domestic infrastructures integral to the whole configuration are generally private market goods, supplied by commercial enterprises, in privately owned houses. As we have seen, the point of connection between the external (state or private monopoly) and domestic infrastructures has been a focus of conflict and continuing redefinition in their economic organisation, and respective liabilities and responsibilities. Moreover, in Delhi, the *absence* of such domestic private infrastructures is a major part of the problem for any generalised public system of tapwater provisioning to the slums. Equally, when public goods domestic tapwater is boiled to make a cup of tea (e.g. Liptons) or instant coffee (e.g. Nescafé), the final consumption good dissolves market with state, in a hybrid distinctive of capitalist political economies.

There are some key points to be examined in turn: the historical and spatial evolution of forms of both public and commodity goods, the dynamic interdependency between public and market goods water, and the many processes of hybridisation of economic goods that are neither commodities nor public goods.

In all our cases, what social or political entity at what scale constituted a 'public' and what form of appropriation or control it exercised over water resources and drinking water provisioning evolved, sometimes quite dramatically, and often through political conflict. Bakker is surely correct in problematising the concept of community control, and the potential conflicts and contradictions between a given community, other communities, and wider-scale publics, including nation states (Bakker, 2010). By adopting an anthropological and IEP conception of property as control (formal and informal; legal, a-legal, and illegal) and by historicising and comparing different evolutionary trajectories of scales of socially collective control over water resources, we can demonstrate the emergence of new scales of 'the public', but also multiple, often heterogeneous publics within the national societal space, as well as the emergence of differential levels of control, hence multi-level publics. Although intimately connected, different 'publics' may be involved in control over water resources and control over water provision.

Thus, in the case of our long duration analysis of London, a royal charter established both the New River Company and its right to appropriate water from the upstream River Lea, an early form of late-feudal state/commercial property rights over water resources, as well as its right to lay pipes across private land. Likewise, rights of abstraction and distribution of water from the River Thames were granted by Parliament to private monopoly producers within the capital. Later still, the municipalisation of water resources witnessed the emergence of public goods water, under some local democratic control, with the formation of London citizens' rights over water and its distribution – not the national public, but localised publics of which London was one amongst many. The 1973 Water Act, put in place by a Conservative but implemented by a Labour Government, for the first time established formal legal status of water resources as a national public good, as well as establishing River Water Authorities as a new scale of public overlaying the level of local government. The national law at this point could be seen as instituting multi-layered levels of control by publics of different scales – quite an unstable arrangement as it turned out. Finally, under privatisation a form of hybrid control emerged, with quite novel non-governmental agencies exercising considerable constraining powers over the monopoly of private water providers in terms of abstraction, distribution, sewage and quality. In relation to the formation of a consumer public, however, rights to drinking water remain highly regionalised, with claims on household disposable income for water varying significantly from one region to another. It cannot be overemphasised how much this process of evolution was politically shaped and, in recent decades, party politically, constituting different scales and layers of public rights over water abstraction and provisioning, often conflictually.

To further underline this point, the sharpest contrast to the evolution of forms of public good in London and the United Kingdom is presented by Taiwan. There, by force of military power of an occupying army, water resources were appropriated and instituted as a national resource. In this case, public goods water has the connotation of an authoritarian state good. During a period of military dictatorship in the years following the Second World War, a national water corporation was formed, the Taiwan Water Corporation, which increasingly extended its control and provision of tapwater to a wider and wider public. It has retained much of this character, even following democratisation in the 1990s. In contrast to the UK, the consumer public, which acquires domestic water through paying water bills, does so as a unified and homogenous public, paying the same tariff for the same tapwater across the whole island.

So economies of drinking water offer a rich empirical resource for conceptualising historical and societal variation of forms of public good. Equally, historical and societal variations in bottled water or tanker water provide an excellent way into exploring varieties of commodity-form. Much of the debate around the commodification of water appears to work within a simplistic view that a commodity is a commodity is a commodity. It is even considered that a full-cost recovery regime of publicly provided water is equivalent to commercial market pricing of a Coca-Cola or Nestlé, just by virtue of the assumption that if a good is paid for with money, then it is a commodity (Swyngedouw, 2004, 35-5, 83; Gleick, 2010). An IEP approach, by contrast, enables significant distinctions to be made between different commodity forms – whether for tomatoes (Harvey et al., 2002), or, in this case water. Commodities are not things but elements within an ongoing process, which involve distinctive forms of markets as points of exchange and distribution, within overall configurations entailing processes of production and consumption. To illustrate this analysis, several contrasting forms of bottled (or commercial tanker) water stand out from our selective cases.

In Germany, a distinctive commodity, highly mineralised sparkling water, is purchased for the price of the water. Consumers buy sparkling water – just the water, not the bottle. In the UK, they buy bottles of water (sparkling or still), and pay the price for so doing. This difference in commodity and market is, in our analysis, tied to a distinctively German cooperation between producers (apart from discount supermarkets) enabling bottles to be recycled through use of an industry standard recyclable bottle, for whose return consumers receive a fee. This kind of producer–retailer cooperation is not shared by the discount supermarkets in Germany, or by the supermarkets in the UK. In the latter, the dominant commodity form, in terms of national market share, is that of own-label supermarket branded bottles of water. The supermarket own-label bottled water in the UK inhabits a market space, alongside the commercially branded bottled waters, quite different from that of Germany, or for that matter, Italy, differentiated both in price terms and processes of qualification (Chapter 3). Consumers purchase bottled water in very different commodity spaces.

If contrasts in commodity form are significant within Europe, they are even sharper in contexts where potable water through pipe infrastructures is unavailable, not trusted, or unreliable. In Europe, bottled water is a supplementary good to tapwater, and in the UK, in terms of the income groups most addicted to its purchase, a luxury good (Chapter 3). In the European context of potable tapwater (however paid for, 'public' or private) bottled water, as a market commodity, ceases to provide for a basic human need. In the context of Mexico City, or indeed Delhi, 'packaged water' occupies an entirely different commodity space, within a very different configuration of production and distribution. If we take the example of Mexico City (Chapter 5), commercial water is predominantly sold as a service, including often the rental of domestic water coolers designed to accommodate 20-litre returnable bottles, and routine deliveries to the household. In short, within this market organisation, the object of purchase is the commodity service of industrially processed water linked to domestic infrastructure, and requiring a different range of consumption work on the part of the purchaser. Although it is of course possible to purchase smaller and portable bottles of water – and statistically more significantly, water in the form of soft drinks – in the Mexico City market and commodity space, from an IEP perspective, the predominant commodity form of home-delivery packaged water is instituted in an entirely different configuration of production, distribution, exchange and consumption, to those characterising the European bottled market economies. Moreover, in Mexico City, bulk packaged water is the primary distributional channel meeting the basic human demand for immediately potable water, in part explaining why Mexico has amongst the highest levels of bottled water and soft drink consumption in the world, in spite of being a developing country. From every aspect, it is a radically different commodity from the luxury supplemental one purchased in the United Kingdom or other European countries. Finally, Mexico City bottled water can be understood only in its interdependence with the other, unevenly accessible, distributional channels of tanker and tapwater, and the fact that these are delivering water for purposes other than drinking. The market space for bottled water is itself conditioned by the economic spaces occupied by tanker water (mostly public) and tapwater, neither of which is potable without substantial treatment, explaining the widespread use of chemical disinfectants for cooking uses when drawing water from domestic storage cisterns.

If drinking water is good for developing an understanding of the evolutionary and societally variable character of both public goods and commodity forms, it is exemplary for exploring the multi-modality and hybridity of capitalist political economies. Multi-modality exhibits a variety of dynamics in different societal contexts, including the Mexico City bottle-tap-tanker interdependency just touched on. The Taiwanese example of water garages provides the supreme instance of multi-modality of capitalist political economies, both technological and monetary. There is literally a direct connection between the public and the private: public tapwater, paid for at the nationally standard domestic rate, feeds commercial operators' filtering equipment delivering grades of water quality to water-pumps, sold as a market commodity to consumers. To be sure, this is an unusual form of directly plumbed-in multi-modality between publicly provided tapwater and commercial enterprises acting as intermediaries (as opposed to the domestic filtering of water). It occurs primarily within a specific context in South West Taiwan, where there has been historically low trust in public tapwater. Moreover, aesthetics of tea drinking as a principal way of consuming water and a valuing of high-tech over natural water purity, underpin consumer demand. In the bottled drinks market, bottled tea far outsells bottled water. So, water garages with their peculiar public–private multi-modality find a particular niche within these Taiwanese economies of water.

Various forms of multi-modality emerge in different societal contexts. The Taiwan example could be seen as a form of a multi-modal chain of provision, another example of which, with a distinctive organisation, was seen amongst Delhi street vendors, whose contractors use treated public tapwater, which is then distributed commercially through street vending carts (Chapter 4). In Mexico City, although much tanker water is both publicly sourced and distributed, private commercial tankers take as their source public treated tapwater (Chapter 5), as do the *tanqueros* of Guayaquil, if in a very different economy of water (Swyngedouw, 2004). In all these cases, the organisation of the chain of provision concerns a division between public and private entities over water processed for consumption, often with a division between public authority processing and related infrastructures, and commercial enterprise distribution. In the case of Mexico City's contracted-out tapwater, both the distribution infrastructure and the water processing is undertaken by the public entity, the commercial enterprise only being responsible for the 'front-end' of consumer services, and, as we have seen, with their commercial identity virtually hidden or disguised (Chapter 5).

At either end of the chain of provision for potable water, two other elements already partly discussed can now be analysed in terms of multi-modality: the interdependence between control over water resources and processes of water production for use; and the interface between a chain of provision and domestic infrastructures. In the case of the UK, we have seen the very gradual emergence of a formal, legally defined, national water resource and the state licensing of abstraction for any use, whether for public tapwater, private tapwater, bottled spring water, or agricultural or industrial uses. In Mexico City and Delhi, the contestation over rights to water resources, and the unstable control over the water resource, serve to highlight the huge significance of establishing a multi-modality between natural resources and their exploitation, whether by either public or commercial entities. In this respect, it is interesting to contrast the case of the Taiwanese state selling licenses to Taiwanese purchasers to legally exploit mountain springs, with the Delhi wealthy illegally extracting groundwater by boreholes in the vacuum of control over putatively public groundwater. In the one case, at least over the short historical term, there is a stable, even growing, economy of water. In the other, a tragedy of the commons. At the other end of the chain of provision, domestic household infrastructures, whether for tap, tanker or bottled water dispensers, illustrate the interdependence between domestic private (non-commercial) property with systems of provision, whether public or commercial. Domestic private property technologies and infrastructures, but also the consumption work of finalising the end product through boiling, making tea, mixing drinks, etc., are critical to the viability and economic sustainability of systems of supply to the household. So, even in cases of totally public control over water resources and provision of potable tapwater, such a public system, within a capitalist political economy, relies entirely on domestic private property, which in turn depends on commercial markets for household technologies and services.

In addition to the multi-modality of chains of provision and their back and front ends, there are also many instances of multi-modal complementarity and competition throughout the phases of a chain of provision. On the one hand, it was observed that at a particular historical instance, the emergence of potable tapwater temporarily destroyed the commercial market for bottled water in some cities in the USA (Chapter 1; Back et al., 1995; Davis and Davis, 1997). On the other, bottled water in contemporary Europe has emerged as a commercial good with a flourishing market complementary to public, privatised, and hybrid state-market tapwater (as discussed above). It is difficult to see bottled water, dedicated for drinking only, as being in competition with tapwater for drinking, and certainly for none of its other uses. In general, in Europe people make tea and coffee with tapwater, perhaps filtered, but not from bottled water. From our drinks diaries (Chapter 3), it is clear that people also drink water as water both tap and bottled, not one at the expense or exclusion of the other. Bottled and tapwater are, in general, non-rivalrous goods.[1] Because water is such a universally consumed good, requiring very different instituted 'publics', the dynamics of competition and complementarity are also quite distinctive. Bottled water producers develop processes of qualification differentiated in general from tapwater and in particular from each other within a space which incorporates both, in order to achieve both complementarity with tapwater and competition within bottled water markets. In the Mexico City or Delhi context, this complementarity is sharper, precisely because that differentiation involves different distribution channels for potable (bottled) and non-potable (tap or tanker) water. Moreover, both in Europe and in the global South, bottled water is in the same market as soft drinks, frequently one product in a portfolio of major bottled drinks manufacturers, such as Coca-Cola, Pepsi Cola, Nestlé or Danone. Bottled water needs to be placed in this wider commodity market space of complementary ways of drinking water out of bottles.

The significance of these emergent varieties of multi-modality is that they highlight the dependency of commercial capitalist enterprises for their very existence on the presence and development of public resources. The soft drinks industry depends on public tapwater as its water source or, in many instances, on the state-licensed abstraction of groundwater, just as much as motor manufacturers depend on public road infrastructures. Water resources exhibit quite distinctive forms of multi-modality, to contrast with but also complement those of knowledge

1 In the very contrasting field of biological knowledge, it was in the balance as to whether the human genome would become a public good or owned by a specialist capitalist enterprise or major pharmaceutical companies, for some critical phases in the early 1990s. As it turned out, the public good databases outcompeted the commercial ventures, but in turn provided critical public resources on which pharmaceutical companies are highly dependent, even with their internal, privately protected, mirror databases (Harvey and McMeekin, 2007). This was a particular configuration of multi-modality.

and the public science base (Harvey and McMeekin, 2007), road infrastructures, or other state-instituted public economies. Analysis of water multi-modality only reinforces a perspective of capitalist political economies as not just driven by capital accumulation, commodity exchange and markets, but as interdependent in a multiplicity of ways with the dynamic growth and innovation of public and hybrid resources. Water flows in multi-modal pathways, revising perspectives on the historical and societal growth of capitalist economies.

Water and quality

Making water safe to drink is certainly a primary imperative in any process of qualification, but rarely is it limited to that in any system of provision, including tapwater. Consider a street vendor in Delhi. The source of the water is first industrially processed by the Delhi Jal Board, with some key quality aspects – safe certainly, but clear and relatively odourless also. It is then purchased by a contractor, filtered again and cooled, and transferred into the regulated standard, licensed, and insulated water vendor trolley. The trolley is then taken by the street vendor to a specified licensed position on the street, its place in the market. The trolley has a number and official branding as a marketing device to guarantee the quality of chilled water to the public. The vendor sells the water in a cup and, although officially prohibited for reasons of hygiene, often with a slice of lemon. For the contractor, the price differential between what they pay to the DJB and what they receive from the sale of a cup of water is the key. For the street vendor, the extra price of including the forbidden lemon is critical for their economic survival (interviews with street vendors, and with officials in the Municipal Delhi Council, August–November, 2012). For the consumer, the taste of lemon-flavoured chilled water, with some official brand assurance for safety, at a point of convenience when thirsty outside the home, is the sought-after bundle of qualities encapsulated in the end product. The process of qualification stretches from the DJB river source, through production, distribution and marketing to consumption: from source to sip.

The comparative analysis of processes of qualification, of both tap and bottled water, displays significant variation, emphasising that there is no singular qualitative substance of 'drinking water'. In Taiwan, although putatively 'safe', especially in cities such as the capital Taipei, tapwater is often filtered, and more universally boiled, before being drunk hot, cold or chilled, either as water or as another beverage, predominantly green tea. In China, people like to drink their plain water when still hot, as a culturally established aesthetic preference, however much choice masks a necessity to make water safe by boiling. Taiwan's water garages offer a wide range of qualities of water for drinking and tea-making, at an equally wide range of prices. There is a similarly broad range of domestic filters, reverse osmosis, UV, carbon, Ph modifying, and so on, either in each individual household, or shared within an apartment block. The IEP source-to-sip approach integrates and emphasises the role of consumption work in defining the quality of water as finally consumed, beyond the market or public infrastructural point of exchange. The process of qualification culminates in a wide range of consumption work practices, societally and cultural variable, of boiling, filtering, chilling, making beverages of all kinds, and so on.

In this book, there has been no systematic comparison between qualification processes of tapwater in Europe. But even from the example of the United Kingdom, there has been a long process of evolution in qualification processes, cutting across private monopoly and public municipality systems of provision. In the nineteenth century, slow-sand filter beds and settling reservoirs achieved a certain quality of biological safety along with reductions in turbidity and odour. In the twentieth century, over half a decade of innovation and development of chlorination resulted in new qualities of tapwater in the second half of the century. The massive programme of eliminating lead piping from external and domestic pipe infrastructures marked a further process of re-qualification, which has subsequently been further elaborated by EU and national regulation. Under privatisation, a distinctive quality regime has been established, with monitoring of over 100 potential contaminants, mineral, chemical and biological. Annual reports highlight consumer complaints and failures to meet quality standards. Nonetheless, these reports reveal considerable variations in qualities of tapwater from one region to another, notably in terms of hardness and softness of water, but also in levels of residual chlorine. People in one area of the country do not consume the same quality of tapwater as those in another, the differences being readily detectable.

Interestingly, the qualification regime assigns a considerable responsibility to the consumer and household, whether to make adjustments to hard and soft water, or to maintain the hygiene security of water tanks and plumbing systems. Brochures are regularly distributed to consumers, both as a quality branding exercise by tapwater providers or the Drinking Water Inspectorate, but also as a device for confirming a division of responsibility for quality between provider and consumer. In this respect, there is a sharp contrast in the organisation of qualification processes between tap and bottled water. In the UK, there are cold water taps fed by water tanks and those under direct mains pressure, only the latter having the highest levels of quality security. Water tanks are an historical legacy from epochs of intermittent and irregular supply, now enshrined in a path dependent on domestic technology of the early industrial leader in piped water supply. Most of continental Europe, a late adopter of piped systems, only has direct mains pressure water from cold taps, so ensuring higher levels of quality security. The linkage between distribution technologies and quality of the water forms an integral aspect of tapwater qualification processes. In a way, the sealed bottling of drinking water provides a linkage between a distribution technology/service and a secure quality of water to mark its difference in qualification processes from tapwater. In tropical or subtropical climates, the significance of that linkage is amplified, as we have seen in Mexico City with its ground-level and roof cisterns, or in Taiwan with its ubiquitous roof tanks. In both these cases, tapwater becomes non-potable through its distribution and domestic storage, and is requalified as potable only when boiled, even though in both instances it was potable at the point of exit from water treatment plants. From this brief discussion, it is clear that societal and historical variations in tapwater provision involve distinctive and evolving qualities of water, beyond any single characteristic of 'safe' or 'potable'.

Comparison of bottled water in Europe also reveals striking differences in qualification processes. In the context of a distinctive tapwater regime, with strong environmental protection of ground and surface water resources, German consumers imbibe highly mineralised and sparkling bottled water. However, as discussed in Chapter 3, European regulation also instituted quality differentiation between mineral and spring water, conflicting with the national market qualification of mineral water established within Germany. There is now a European space of regulatory qualification superimposed on national societal regimes of qualification. The Perrier green bottle – an historical institution – is a market device, but it comes at a price, a dimension missing from a Callonian analysis. Different national bottled water markets place the Perrier brand in different quality and price market spaces. In the UK, the branding of supermarket own-label provides a strong contrast in bottled design, display, shelf-location, and price with manufacturer-branded bottles such as Perrier. In Germany, as we have seen, there is a divided market between dedicated retail outlets and the discount supermarkets. In Italy, partly promoted by supermarket market strategy, loyalty and emphasis on locally sourced bottled waters compete with the multinational brands. A distinctive Italian culture of quality, whether for food or water, valuing products of defined geographical origin, shapes the quality–price matrix (Kjaernes et al., Chapter 8, 2007; Allaire in Harvey et al., 2004). In short, branding as market device is always inscribed within a quality–price matrix, varying widely from one societal market organisation to another, and in different configurations of production, distribution, exchange and consumption. The radical difference between qualification processes of bottled and tapwater, and the fact that neither can be reduced to a universal, asocial, or ahistorical quality outcome of 'safe to drink' enables us to better understand the huge variation in the consumption practices of the drinking of water.

Society–nature variation and sociogenesis of sustainability crises

So far in this chapter, we have been considering historical and societal variation in transitions, forms of economic good, multi-modality of capitalist political economies, and qualification processes. In each of the chapters on London, Mexico City, Delhi and Taiwan, historical and societal development of economies of water were placed in distinctive environments of water resources. In the case of London, beyond the obvious point that the city was built on the continuous, but also tidal, flow of the Thames, its distinctiveness amongst big cities even within the UK was the sufficiency of the Thames river basin, without recourse to dams and long-distance transport of water. The switch from ground to surface water as a principal water resource occurred early in its metropolitan expansion, and even though it was marked by successive and severe droughts throughout the nineteenth and twentieth centuries, post hoc expansion of reservoirs and water treatment capacities largely met growing demand. Inequalities in rights to tapwater, in terms of reliability and price persisted until municipalisation (Trentmann and Taylor, 2005), but these were nothing compared with the scale and depth of inequalities manifest in contemporary Delhi. Delhi, also adventitiously situated on a major river, has progressively and controversially extended its reach over water resources to dams built hundreds of kilometres from the city. Yet, the severe water poverty experienced by millions of its citizens is due not to environmental scarcity, but rather to political inertia, strategic exclusion, and the chaotic agglomeration of an ever-expanding megacity. If and when tapwater was to become as universally available in Delhi as it is to citizens of London, then no doubt greater claims on more distant resources would be entailed, indubitably in competition and conflict with other societal claims. By contrast, Mexico City's water resource environment was fundamentally altered by an act of war, destroying the Aztec hydraulic management forever. Yet, in so doing, a valley-bottom space was created for the eventual emergence of a megacity, consequentially reliant on obtaining water from ever greater distances, at the ever greater energy costs of pumping water up to a height of 2000 metres. Taiwan, frequently deluged by typhoons and endowed if anything with a superabundance of water, also switched from a dominance of groundwater wells to cities fed by mountain rivers on their short flow from mountain spine to the sea, but only once these had been violently appropriated and subsequently controlled by the systematic building of dams under two successive authoritarian regimes.

It would be quite mistaken to characterise these diverse trajectories as societal adaptations to environmental water resources, let alone as social construction of abundances and scarcities (Kaika, 2003). Although in interaction, the dynamics of societal and environmental or biological (human and non-human) processes have their own independence. As we saw in Evans' analysis of cholera in Hamburg and its contrast with neighbouring Altona (Evans, 1987), political, economic and ideological processes eventuated in a catastrophic epidemic only in interaction with the biological characteristics of humans and bacilli, in the seasonal environmental conditions of the River Elbe (see Chapter 1). Similarly, Davis has argued that the oceanic phenomena of the El Niño-Southern Oscillation, with their own independent climatic and environmental dynamics, interacted with the colonial political regimes, world trade and economic transformations of Indian society, to induce the devastating famines of the Victorian epoch (Davis, 2002).

The interactions involved in our cases are asymmetric, driven one-sidedly by societal (especially urban) growth, exploitative of their local water environments. Without the hindsight of history, London appears accidentally fortunate in its water resource endowments, and relatively conflict free. The siting of Beijing in the dry North East of China, again in hindsight, by contrast appears accidentally misplaced – now requiring the gigantic engineering feat of the South-North Water Transfer Project, transporting water from the Himalayan foothills for 1100 kilometres (Zheng et al., 2010; Rong, 2011; Wang and Huang, 2013). The dynamics of Beijing's growth occurred almost despite its known water environment of relative scarcity. São Paulo, however, sited in an environment of relative water abundance for the last several centuries, suddenly, and probably as a consequence of climate change, finds itself in 2015 accidentally and unpredictably out-of-kilter with that environment after a seasonal shift in rainy seasons and a succession of severe droughts. It now faces major energy crises from its stranded hydroelectric power stations as well as a crisis of water supply to support its 20 million citizens (*Financial Times,* February 11, 2015). Finally, the example of Mexico City stands almost as a counterfactual, as a relatively abundant water resource environment collaterally damaged by inadvertent human action, only to release space for the growth of a megacity almost uniquely unsuccoured by a major umbilical river.

But above all the asymmetric and frictional interactions between cities and their national water environmental resources is manifest in the distinctive sustainability crises that have inflected each of their varied historical trajectories. Water provides a particularly useful and also distinctive way in to the analysis of sustainability crises as sociogenic, rather than anthropogenic. Crises are consequential upon particular societal interactions and trajectories in relation to their particular environmental consequences (Harvey, 2014a). If crises are generated by historical societies, and not by a generic agglomerated Anthropogen, they also have particular societal impacts, affecting some sectors of societies, regions, or nations differently than others. So they are both sociogenic and sociotopic. The example of São Paulo's drought and water crises is an example of sociotopic impact probably induced climate change, which has societally been generated in very different ways: Brazil has one of the lowest proportions of its greenhouse gas emissions from energy extraction, industry, and transport in the world, and one of the highest from agriculture and deforestation. For climate change, the disjuncture between the sociogenic generation and the sociotopic impacts can be quite extreme – witness the effects of rising sea levels on low-lying tropical and sub-tropical islands. In our cases of water sustainability crises, the linkages between generation and impact are much more proximate: directly water borne within the immediate environment. Even Bazelgette's barges transporting London's sewage into the North Sea for a hundred years had relatively proximate sociotopical impacts on the coastal beaches of its periphery, despite the effort to create distance (Chapter 2). But it is precisely the distinctive relation between sociogenic generation and sociotopic impact of water sustainability crises that can make a contribution to developing an analysis of sustainability crises more widely.

There is a surface similarity between the remark made by Thomas Cubitt in 1840 describing the Thames in London as 'one big cesspool' and that of the Chief Engineer of the Delhi Jal Board in 2012 designating the river Yamuna as 'a sewer', a surface similarity replicated in Chicago in the late nineteenth century, Hamburg, or the Netherlands. In Chicago, the flow of sewage was switched in 1900 from going into Lake Michigan, the city's water source, to the Mississippi, so that St Louis enjoyed the sociotopic impact (Salzman, 2014). But a comparison between the mid-nineteenth century sewage sustainability crisis in London and the twenty-first century one in Delhi reveals the very different societal dynamics, defining the sociogenic rather than generic nature of the crises. In London, the Great Stink arose largely as a consequence of the rapid growth of piped water infrastructures together with the spread of the water closet, a technological change then combined with a legal and political one, that switched proscribing to prescribing the flow of sewage into the Thames by the sewage system. The configurational analysis of this change stressed the interdependence between water provision by private monopolies and sewage management by public authorities. Remarkably, although the main sustainability risk was water-borne disease with the spread of cholera through Northern Europe from the 1830s, both the then current theory of miasmatic transmission of disease and the political impact of the Great Stink on the Houses of Parliament triggered the major configurational change and the construction of an intercept sewer system. One of the key aspects of the crisis was the tidal nature of the Thames in London: what flows out tends to flow back in, one of the parameters of the sociotopic impact. It was indeed the 'Stink', not a devastating cholera epidemic, which triggered the change, unlike Hamburg.

By contrast, in contemporary Delhi, although the scientific understanding of water-borne diseases is now paradigmatic, the pollution of the Yamuna is primarily a consequence of the *lack* of connection of the vast majority of the population to a sewer system, and the consequential unregulated discharge of all kinds of pollutants into the river, directly and through open drains and gulleys. In our analysis, this contrast is intimately linked to the chaotic agglomeration of Delhi at a pace and scale unparalleled in early nineteenth-century London. However, this process of urban growth is combined with the politics of strategic informality and exclusion: the pollution of the Yamuna evokes a huge shrug of Delhi's huge and divided bureaucracy's shoulders suffocating in a miasma of political inertia. Moreover, the Yamuna in Delhi is non-tidal, so much of the sociotopical impact is transferred downstream of the city, only in part to return in the form of contaminated foodstuffs grown in the affected agricultural riverine land (Chapter 4; Toxics Link, 2014). In terms of flows, and their IEP configurational analysis, Delhi draws on often distant upstream water resources, and discharges effluents downstream, passing through Delhi's notorious belly. Mexico City provides a further instance of societal variation underlying sociogenic-sociotopic sustainability crises related to sewage. There, a relatively high proportion of households are connected to the sewage system, but the effluents are efficiently released untreated into the river systems at some distance and drop from the height of the mountain basin. So, the sociotopic impact is extreme for agricultural populations especially outside and below the city, only returning in part, as in Delhi, in the form of fertilised-contaminated, outsized fruit and vegetables. In Mexico City's distinctive configuration, consumers respond with the widespread addition of chemical bactericides to their cooking water (Chapter 5).

This schematic comparison of sewage-related sustainability crises only partially encompasses their sociogenic character and dynamics. In the case of Delhi and Mexico City there are multiple and interconnected sustainability crises, requiring a holistic analysis. In Delhi, lack of reliable network connectivity, of both tapwater and sewage, has resulted in extensive groundwater contamination. As we saw, intermittent tapwater pressures and supply result in leaks-in as well as leaks-out, so rendering the tapwater unpotable without further treatment. The inadequacies of the tapwater system furthermore induce wealthier households, some two million of them, to dig private domestic and illegal borewells, resulting in over-abstraction and unsustainability of groundwater resources. The sewage sustainability crisis is dynamically related to the groundwater over-abstraction crisis. In Mexico City, the relative deprivation of surface water in the location of the city has driven overdependence on groundwater abstraction, leading to the very distinctive and recurrent sustainability risk of subsidences, also damaging the tapwater infrastructures. As a consequence, tapwater in Mexico City is not a sustainable source of potable water. Moreover, reliant now on distant sources such as the Cutzamala river basin for its water provisioning, the city is subject to regular flooding from surface water descending, unharnessed, from the surrounding mountains, carrying with it further contamination as it descends from the city to the valleys below. In Mexico City as in Delhi, the sustainability crisis is a dynamically combined set of interactions between socio-economies of water and sewage, within the particular characteristics of its environment and seasonal patterns of rainfall.

In concluding this section, society-nature interactions can be seen both as asymmetrically driven, and as involving social and natural processes each with their own independent dynamics. This was demonstrated in terms both of the dynamics of growth and development of societies and cities in relation to their water resource environments, and of the sociogenic/sociotopic sustainability crises. Whether a river is tidal or non-tidal at a particular location matters. So does the historical development of the societal organisation of water drawn from it, and flowing back into it. The dynamic relation between generation (sociogenic) and impact (sociotopic) of a sustainability crisis can be socially modified, certainly, but only within natural environmental parameters and dynamics. In this analytical perspective, society-nature interactions represent a critical dimension in understanding the historical and societal variations of socio-economies of water.

Water and variation

So far in this chapter, the focus has been on the use of comparison in analysing variations in how water is made drinkable and how it is drunk in a variety of historical, societal and environmental contexts. In this final section, the focus is reflected back on itself to suggest how the analysis of drinking water might contribute to a more general social scientific understanding. The significance of water as an object of social scientific analysis is that it embodies and exemplifies the three main parameters of variation: societal organisation, historical change and society-nature interaction. Given that drinking water is an existential condition of society, and that its organisation is always social, water is always at the fulcrum of the three dynamics of variation. It is fruitful to analyse the social organisation of water as a condition for human social existence, rather than in terms of satisfaction of individual asocial need – bracketing-off the biological from the societal. The analysis of drinking water provides a useful model of social scientific explanation in terms of a three-way *variation-in-combination*. In many analyses of varieties of capitalism (e.g. Esping-Anderson, 1990; Whitley, 1999; Hall and Soskice, 2001; Boyer, 2005), dealing with welfare states, state–market relations, forms of governance, it has been possible to bracket-off variations in society-nature interactions, even when purporting to address the comparative advantage of different political economies. In the case of water, that bracketing-off makes no sense, but to the contrary, those interactions in combination with societal and historical variation are critical for understanding the development of economies of water and their sociogenic/sociotopic sustainability crises.

By exploring three-way variation-in-combination through the refracting surface of water in common with other studies of the social life of economic objects (Appadurai, 1986), such as sugar (Mintz, 1985), cod (Kurlansky, 1999), or the tomato (Harvey et al., 2002), there is a particular analytical emphasis on open-ended historical change, eluding any pre-emptive classification in terms of societal variation. In the comparative analysis presented here, there were no absolute beginnings, and no fixed ends to the exploration of historical changes in drinking water in any societal context (Harvey, 2014b). The municipal hydraulic model, for example, emerged in history, unevenly and in different societal forms, and has not arrived at any end of history. Socio-economic forms of urban tapwater provisioning are continuously evolving. In this book, no attempt has been made to classify, or taxonomically freeze, societal water economies into boxes of market-environmentalism, state corporatism, communitarianism, or any other fixing classification. The analysis of historical variation privileges open-ended process over outcomes at any temporal fix.

And of course, water is not alone in exemplifying the three-way variation-in-combination. Land resources (especially for food), energy resources, and a variety of materials resources are entering into the forefront of analysis as societies and polities confront the increasingly pressing challenges of climate change and the finitudes of planetary resources in their environmental locations (Urry, 2010, 2011, 2014). It has been argued that the 'great divergence' in the world between industrialising and non-industrialising societies and economies at the time of the industrial revolution occurred through a combination of the opening up of new land resources for food and clothing in the New World, and exploitation of coal in the Old (Pomeranz, 2000). Now that virtually the whole world is on various trajectories of capitalist or market socialist growth, planetary resources and their location, as well as climate change, have thrust society-nature interactions into historically unprecedented significance. The relation between society, economy and nature is now centre-field for social scientific analysis. For social sciences, three-way variational analysis is no longer optional.

Karl Polanyi, in analysing the emergence of industrial capitalism, wrote of the 'shifting place of economy in society', with the economy dominating the polity (Polanyi, 1944). He also wrote of the destructive potential to land (to which we may easily add water) and human wellbeing, if the market economy expanded without political or social restriction. But his conception of society-nature interaction (biological and environmental) was relatively underdeveloped, and his implication that society would correct itself, swinging from unregulated to state regulated capitalism in what he denoted 'the double movement' is controverted by many of the historically acute sustainability crises including those analysed in this book. The dire examples of Delhi and the Yamuna, Mexico City and its environs stand as evidence to subvert any optimism for a socially resilient double movement. O'Connor, drawing heavily on Polanyi in his greening of Marxism (O'Connor, 1998), argued that only interventions of the state, social movements and politics, offered a possibility of mediating between the capitalist economy and nature. Our analysis of the three-way variation-in-combination in the evolution of water economies has also highlighted the central significance of politics in mitigating or intensifying sustainability crises, including ideological and political paralysis or tunnel vision.

So, we need to extend Polanyi's analysis of the 'shifting place of economy in society' with a parallel emphasis on the 'shifting place of society in nature' in the contemporary historical context of climate change and finite planetary resources. Drinking water has a big story to tell. On the one hand, sustainability crises are generated sociogenically by particular societies with their particular historical trajectories in relation to particular environmental resources. On the other hand, sustainability crises impact sociotopically often on the poorest and most deprived, often at a distance or even disconnected from their generation. As with water, the crises of climate change and rights to, competition for, and conflicts over, finite resources can only be met by politics of many levels, from the local, through the societal to the international, recognising the varied dynamics of their generation and of their impacts across the world.

Bibliography

Acreman, M., Dunbar, M., Hannaford, J., Mountford, O., Wood, P., Holmes, N. and Crookall, D. 2008. 'Developing environmental standards for abstractions from UK rivers to implement the EU Water Framework Directive', *Hydrological Sciences Journal*, 53, 6, 1105–20.

Agrawal, A. 2003. 'Sustainable governance of common pool resources: context, method and politics', *Annual Review of Anthropology*, 32, 243–62.

Agua Rios Y Pueblos. 2014. www.aguariosypueblos.org/ Accessed 21.02.2014.

Alaszewski, A. 2006. *Using Diaries for Social Research*. London: Sage.

Ali, S. 2003. *Environmental Situation of Slums in India*. Uppal Publishing House: Delhi.

Allaire, G. 2004. 'Quality in economics: a cognitive perspective', in Harvey, M., McMeekin, A. and Warde, A. (eds) *Qualities of Food*. Manchester University Press: Manchester.

Anand, N. 2012. 'Municipal disconnect: on abject water and its urban infrastructures', *Ethnography*, 13, 4, 487–509.

Appadurai, A. (ed.) 1986. *The Social Life of Things. Commodities in Cultural Perspective.* Cambridge University Press. Cambridge.

Appadurai, A. 2001. 'Deep democracy: Urban governmentality and the horizon of politics', *Environment and Urbanization*, 13, 2, 23–43.

Asthana, V. 2009. *Water Policy Processes in India: Discourses of Power and Resistance*. Routledge: Delhi.

Austen, R.A. and Smith, W.D. 1992. 'Private tooth decay as a public economic virtue: the slave-sugar triangle, consumerism, and European industrialisation', in Inikori, J.E. and Engerman, S.L. (eds) *The Atlantic Slave Trade: effects on economies, societies and people in Africa, the Americas and Europe*. Duke University Press: Durham, NC. 183–204.

Back, W., Landa, E.R. and Meeks, L. 1995. 'Bottled water, spas, and early years of water chemistry', *Ground Water,* 33, 4, 605–14.

Bakker, K. 2002. 'From state to market?: water *mercantilaçion* in Spain', *Environment and Planning A*, 34, 767–90.

Bakker, K. 2003a. *An Uncooperative Commodity: Privatizing water in England and Wales.* Oxford University Press: Oxford.

Bakker, K. 2003b. 'From public to private to…..mutual? Restructuring water supply governance in England and Wales', *Geoforum*, 34, 359–74.

Bakker, K. 2007. The "commons" versus the "commodity": Alter-globalization, anti-privatization and the human right to water in the global South. *Antipode*, 39, 3, 430–55.

Bakker, K. 2010. *Privatizing Water: Governance Failure and the World's Urban Water Crisis*. Cornell University Press: Ithaca.

Bakker, K. and Bridge, G. 2006. 'Material worlds? Resource geographies and the "matter of nature"', *Progress in Human Geography* 30, 1, 5–27.

Banerji, M. 2005. 'Provision of basic services in the slums and resettlement colonies of Delhi', *Institute of Social Studies Trust*, Delhi, 1–18.

Barlow, M. and Clarke, T. 2002. *Blue Gold. The fight to stop the corporate theft of the world's water*. Earthscan: Oxford.

Barquera, S., Hernandez-Barrera, L., Tolentino, M.L., Espinosa, J., Ng, S.W., Rivera, J.A. and Popkin, B.M. 2008. 'Energy intake from beverages is increasing among Mexican adolescents and adults'. *The Journal of Nutrition*, 138, 12, 2454–61.

Barrett, R.E. & Whyte, M.K. 1982. 'Dependency theory and Taiwan: Analysis of a deviant case' *American Journal of Sociology*, 1064–1089.

Bateman, I.J., Langford, I.H., Nishikawa, N. and Lake, I. 2000. 'The Axford debate revisited: A case study illustrating different approaches to the aggregation of benefits data', *Journal of Environmental Planning and Management*, 43, 2, 291–302.

Baud, I., Sridharan, N. and Pfeffer, K. 2008. 'Mapping urban poverty for local governance in an Indian mega-city: the case of Delhi', *Urban Studies*, 45, 7, 1385–1412.

Bhagat, R.B. and Mohanty, S. 2009. 'Emerging pattern of urbanisation and the contribution of migration in urban growth in India', *Asian Population Studies*, 5, 1, 5–20.

Bhan, G. 2009. '"This is no longer the city I once knew". Evictions, the urban poor and the right to the city in millennial Delhi', *Environment and Urbanization*, 21, 1, 127–42.

Boo, K. 2012. *Behind the Beautiful Forevers: Life, death and hope in a Mumbai slum*. Portobello Books: London.

Borroto, R.J. 2000. 'Geographical patterns of cholera in Mexico, 1991–1996', *International Journal of Epidemiology*, 29, 4, 764–72.

Botto, S. 2009. 'Tap vs bottled water in a footprint integrated approach', *Nature Precedings*, http://precedings.nature.com/documents/3407/version/1 Accessed 03.05.2012.

Bourdieu, P. 1984. *Distinction: A social critique of the judgement of taste*. Harvard University Press: Cambridge, MA.

Boyd, G.R., Tarbet, N.K., Kirmeyer, G., Murphy, B.M., Serpente, R.F., and Zammit, M. 2001. 'Selecting lead pipe rehabilitation and replacement technologies', *Journal of the American Water Works Association*, 93, 7, 74–87.

Boyer, R. 2005. 'How and why capitalisms differ', *Economy and Society*, 34, 4, 509–57.

Brei, V. and Böhm, S. 2011. 'Corporate social responsibility as cultural meaning management: a critique of the marketing of "ethical" bottled water', *Business Ethics: A European Review*, 20, 3, 233–52.

Broich, J. 2007. 'Engineering the Empire: British water supply systems and colonial societies, 1850–1900', *Journal of British Studies*, 46, 2, 346–65.

Broich, J. 2013. *London. Water and the Making of the Modern City*. University of Pittsburgh Press: Pittsburgh, PA.

Browder, G.J. and Xie, S. 2007. *Stepping Up. Improving the Performance of China's Urban Water Utilities*. World Bank: New York.

Buijzen, M., Schuurman, J., and Bomhof, E. 2008. 'Associations between children's television advertising exposure and their food consumption patterns: A household diary–survey study', *Appetite*, 50, 2, 231–39.

Callon, M. 1986. 'Some elements of a sociology of translation: domestication of scallops and the fishermen of St Brieuc Bay', in Law, J. (ed.) *Power, Action, and Belief: A New Sociology of Knowledge*. Routledge: London.

Callon, M., Méadal, C., and Rabeharisoa, V. 2002. 'The economy of qualities', *Economy and Society*, 31, 2, 194–217.

Castano, I. 2012. Mexico's Water War. *Forbes*. Retrieved from www.forbes.com/sites/ivancastano/2012/02/22/mexicos-water-war/ Accessed 18.06.2013.

Castro, J.E., Kaika, M. and Swyngedouw, E. 2003. 'London: structural continuities and institutional change in water management', *European Planning Studies*, 11, 3, 283–98.

Chakrabarti, P.G.D. 2001. *Delhi's Ongoing Debate on Informal Settlements and Work Places – Issues of Environmental Jurisprudence*. Paper presented at the International Workshop on Coping with Informality and Illegality in Human Settlements in Developing Cities.

Chan, S. 2003. The modernity and national character in the management of a colonial city, Master's Dissertation, Taiwan: Chung Yuan Christian University. (Text in Chinese). www.klci.design.stu.edu.tw/databases/G_9_15.pdf Accessed 21.11.2014.

Chan, Y.N. (ed.) 2006 *Taiwan's Tap Water*, Taiwan Tap Water Association: Taipei. (Text in Chinese).

Chatterjee, P. 2006. *The Politics of the Governed: Reflections on Popular Politics in Most of the World*. Columbia University Press: New York.

Chen, H.T. 2001. Taiwan Water Business operation in the Japanese era: The example of the JiaNan irrigation water system〈日治時期臺灣水利事業的建立與運作：以嘉南大圳為例〉. Fu-Jen Historical Journal《輔仁歷史學報, 12, 117–52. (Text in Chinese).

Cheng, H. 2009. The rules about building during the Japanese occupation in Taiwan, in the Encyclopedia of Taiwan, Ministry of Culture. (Text in Chinese), http://taiwanpedia.culture.tw/web/content?ID=4877# Accessed 21.09.2013.

Chevalier, W.S. 1953. *London's Water Supply, 1903-1953*. The Metropolitan Water Board, Staples Press: London.

Chin, Hsien-Yu. 1998. 'Colonial Medical Police and Postcolonial Medical Surveillance Systems in Taiwan, 1895–1950s', Beyond Joseph Needham: Science, Technology, and Medicine in East and Southeast Asia. *Osiris*, 2nd Series, Vol. 13, 326–38.

Cleaver, F. 2002. 'Reinventing Institutions: Bricolage and the Social Embeddedness of Natural Resource Management', *The European Journal of Development Research* 14, 2, 11–30.

Cleaver, F. 2012. *Development through Bricolage: Rethinking institutions for natural resource management*. Routledge: London.

Cohen, M.P., and Reynoso, A.E.G. 2006. 'Del agua amenazante al agua amenazada. Cambios en las representaciones sociales de los problemas del agua en el Valle de México', in Urbina, J. and Martínez, J. (eds) *Más allá del cambio climático: Las dimensiones psicosociales del cambio ambiental global* (p. 287). Mexico DF: Instituto Nacional de Ecología. Retrieved from http://books.google.com/books?hl=en&lr=&id=vvUoRldp2nkC&pgis=1 Accessed 18.06.2015.

Conagua. 2011. *Estadísticas del agua en México* , edición 2011. Mexico City.

Conagua. 2012. *Sistema Nacional de Tarifas*. www.conagua.gob.mx/Tarifas/paginas/Principal.aspx Accessed 16.08.2012.

Crook, T. 2006. 'Schools for the moral training of the people': public baths, Liberalism and the promotion of cleanliness in Victorian Britain', *European Review of History*, 13, 1, 21–47.

Daga, S. 2002. Private Supply of Water in Delhi, *Working Paper No. 58 – State, Market & Economy*. Centre for Civil Society: Delhi.

Davis, M. 2002. *Late Victorian Holocausts. El Niño famines and the making of the Third World*. Verso: London.

Davis, M. 2007. *Planet of Slums*. Verso: London.

Davis, S.N. and Davis, A.G. 1997. 'Saratoga Springs and early hydrogeochemistry in the United States', *Ground Water*, 35, 2, 347–56.

DEFRA, 2010. 'River Water Quality Indicators for Sustainable Development – Annual Report.' https://www.gov.uk/government/uploads/system/uploads/attachment_data/file/141697/rwq-ind-sus-2009-resultsv2.pdf Accessed 21.01.2015.

Delgado-Ramos, C.G. 2014. 'Water and the political ecology of urban metabolism: the case of Mexico City', *Journal of Political Ecology*, 22, 98–114.

Delhi Statistical Handbook. 2010. Government of NCT of Delhi: Delhi.

Demarest, A. 2004. *Ancient Maya: the rise and fall of a rainforest civilization.* Cambridge University Press: Cambridge.

Department of Health, 1995. *The history of public health development of Taiwan area,* Vol. 1–5, Taipei: Executive Yuan. (Text in Chinese).

Diduch, M., Polkowska, Z., and Namiesnik, J. 2011. 'Chemical quality of bottled water: a review', *Journal of Food Science*, 76, 9, R178–96.

Drinking Water Inspectorate. 2012. *Securing Safe, Clean Drinking Water for All.* Drinking Water Inspectorate: London.

Drinking Water Inspectorate. 2013. *Drinking Water 2012. Chief Inspector's Report* Drinking Water Inspectorate: London.

D'Souza, R. 2006. 'Water in British India: the making of "colonial hydrology"', *History Compass*, 4, 4, 621–28.

Dun, Y. 2005. *The Empire and the Toilet.* Taiwan History Materials Publisher: Taipei. (Text in Chinese).

Dupont, V. 2004. 'Socio-spatial differentiation and residential segregation in Delhi: A question of scale?' *Geoforum,* 35, 157–75.

Dutta, V., Chander, S. and Srivastava, L. 2005. 'Public support for water supply improvements: Empirical evidence from unplanned settlements of Delhi, India', *The Journal of Environment and Development,* 14, 4, 439–62.

Dutta, V. and Verma, N. 2009. 'Water pricing, affordability and public choice: An economic assessment from a large Indian metropolis', *International Journal of Economics and Finance,* 1, 2, 238–52.

Environment Agency. 2008. *Abstracting Water: a Guide to Getting Your Licence.* Environment Agency: Bristol.

Esping-Anderson, G. 1990. *The Three Worlds of Welfare Capitalism.* Polity: Cambridge.

Euromonitor. 2009. *Bottled Water – Germany.* Euromonitor International: Country Sector Briefing May 2009.

Euromonitor. 2009. *Bottled Water – Italy.* Euromonitor International: Country Sector Briefing May 2009.

Euromonitor. 2009. *Bottled Water – United Kingdom.* Euromonitor International: Country Sector Briefing May 2009.

Evans, A. and Miele, M. 2012. 'Between food and flesh: how animals are made to matter (and not matter) within food consumption practices', *Environment and Planning D: Society and Space*, 30, 2, 298–314.

Evans, R.J. 1987. *Death in Hamburg. Society and Politics in the Cholera Years, 1830–1910.* Penguin: London.

European Union. 2000. *The Water Framework Directive.* http://eur-lex.europa.eu/legal-content/EN/TXT/?uri=CELEX:32000L0060 Accessed 08.02.2012.

Executive Yuan, 1962a. The secretary meeting notes, First division no. 3166. Taipei: Executive Yuan Archives.

Executive Yuan, 1962. The secretary meeting notes, First division no. 3248. Taipei: Executive Yuan Archives.

Executive Yuan, 1962b. The secretary meeting notes, First division no. 3241. Taipei: Executive Yuan Archives.

Executive Yuan, 1962c. The secretary meeting notes, First division no. 3363. Taipei: Executive Yuan Archives.

French, S.A., Lin, B.H., & Guthrie, J.F. 2003. 'National trends in soft drink consumption among children and adolescents age 6 to 17 years: prevalence, amounts, and sources, 1977/1978 to 1994/1998' *Journal of the American Dietetic Association*, 103, 10, 1326–31.

Fu, K. and Ho, Y. 2012. Urban Planning Story: The project of Taiwan history digital archives. (Text in Chinese). www2.ndap.org.tw/newsletter06/news/read_news.php?nid=505 Accessed 21.10.2013.

Fullalove, S.K. (ed.) 1994. Thames Water Ring Main. Civil Engineering Special Issue. Thomas Telford Ltd.

Geels F.W. 2002. 'Technological transitions as evolutionary reconfiguration processes: a multi-level perspective and case study', *Research Policy*, 31, 1257–74.

Geels, F.W. 2004. 'From sectoral systems of innovation to socio-technical systems: Insights about dynamics and change from sociology and institutional theory', *Research Policy*, 33, 6, 897-920.

Geels, F.W. 2006. 'The hygienic transition from cesspools to sewer systems (1840–1930): the dynamics of regime transformation', *Research Policy*, 35, 7 1069–82.

Geels, F.W. and Schot, J. 2007. 'Typology of sociotechnical transition pathways', *Research Policy*, 36, 399–417.

George, L.D. 1994. 'Uses of spring water', *Environmental Geology*, 24, 125–32.

Gleick, P.H. 2004. 'The Millennium Development Goals for Water: Crucial objectives. Inadequate Commitments, Gleick, P.H., (ed.) *The World's Water, 2004–2005*, Island Press: Washington, 1–15.

Gleick, P.H. 2010. *Bottled and Sold: The Story Behind our Obsession with Bottled Water*. Island Press: Washington DC.

Gleick, P.H. 2011. *The World's Water. Volume 7*. Island Press: Washington DC.

Gleick, P.H. and Cooley, R.S. 2009. 'Energy implications of bottled water', *Environmental Research Letters*, 4, 1.

Glennie, P.D. and Thrift, N.J. 1993. 'Modern consumption: theorising commodities and consumers', *Environment and Planning D, Society and Space*, 11, 603–606.

Glucksmann, M. 1990. *Women Assemble. Women workers and the new industries in inter-war Britain.* Routledge. London

Glucksmann, M. 2000. *Cottons and Casuals. The gendered organisation of labour in time and space.* Sociologypress. British Sociological Association. Durham.

Glucksmann, M. 2009. 'Formations, connections and divisions of labour', *Sociology*, 43, 878–95.

Glucksmann, M. 2013. 'Working to consume: consumers as the missing link in the division of labour.' Centre for Research in Economic Sociology and Innovation (CRESI) Working Paper 2013-03, University of Essex, http://repository.essex.ac.uk/7538/ Accessed 16.12.2014.

Godelier, M. 1970. *Sur les sociétés précapitalistes*. Paris: Edition Sociales.

Gómez, A. 2010. Estado y Política Hidráulica en México. El Caso de los Indígenas Mazahuas. *redissa.hostei.com*. Retrieved from http://redissa.hostei.com/rissa/Gomez_Fuentes_Anahi.pdf Accessed 06.06.2013.

Goubert, J-P. 1989. *The Conquest of Water: The Advent of Health in the Industrial Age*. Polity: Cambridge.

Graham, S. and Marvin, S. 2001. *Splintering Urbanism: Networked Infrastructures, Technological Mobilities, and the Urban Condition*. Routledge: London.

Graham-Leigh, J. 2000. *London's Water Wars: The Competition for London's Water Supply in The Nineteenth Century*. Francis Boutle: London.
Hall, P.A. and Soskice, D. 2001. *Varieties of Capitalism: The Institutional Foundations of Comparative Advantage.* Oxford University Press: Oxford.
Halliday, S. 2001. *The Great Stink of London: Sir Joseph Bazalgette and the Cleansing of the Victorian Metropolis*. The History Press: Stroud.
Hamlin, C. 1992 'Edwin Chadwick and the Engineers, 1842–1854: systems and antisystems in the pipe-and-brick sewers war', *Technology and Culture,* 33, 4, 680–709.
Hanley, J.G. 2006. 'The metropolitan commissioners of sewers and the law, 1812–1847', *Urban History*, 33, 3, 350–68.
Hartley, D. 1964. *Water in England.* MacDonald and Jane: London.
Harvey, M. 2007a. 'Working with and beyond Polanyian perspectives', in Harvey, M. Ramlogan, R. and Randles, S. (eds) *Karl Polanyi: New Perspectives on the Place of the Economy in Society.* Manchester University Press: Manchester.
Harvey, M. 2007b. 'Instituting economic processes in society', in Harvey, M., Ramlogan, R. and Randles, S. (eds) *Karl Polanyi: New Perspectives on the Place of the Economy in Society.* Manchester University Press: Manchester.
Harvey, M., Quilley, S. and Beynon, H. 2002. *Exploring the Tomato: Transformations in Nature, Economy and Society.* Edward Elgar: Cheltenham.
Harvey, M., McMeekin, A. and Warde, A. (eds) 2004. *Qualities of Food.* Manchester University Press: Manchester.
Harvey, M. and McMeekin, A. 2007. *Public or Private Economies of Knowledge? Turbulence in the Biological Sciences*. Edward Elgar: Cheltenham.
Harvey, M. 2014a. 'On the horns of the food-energy-climate change trilemma: towards a socio-economic analysis', *Theory, Culture and Society*, 31, 5, 155–82.
Harvey, M. 2014b. 'Comparing comparing: exercises in stretching–concepts', *Anthropology of Food,* S10. http://aof.revues.org/7624 Accessed 30.09.2014.
Hassan, J. 1985. 'The Growth and Impact of the British Water Industry in the Nineteenth Century', *Economic History Review*, 38, 4, 531–47.
Hassan, J. 1998. *A History of Water in Modern England and Wales*. Manchester University Press: Manchester.
Haug, G.H., Günther, D., Peterson, L.C., Sigman, D.M., Hughen, K.A. and Aeschlimann, B. 2003. 'Climate and the collapse of Maya Civilization', *Science*, 299, 1731–35.
Hawkins, G. 2009. 'The politics of bottled water: assembling bottled water as brand, waste and oil', *Journal of Cultural Economy*, 2, 1–2, 183–195.
Hayes, C.R., Bates, A.J., Goodman, A.H., Vinson, J.P. and Sadler, T.P. 1997. 'Meeting standards for lead in drinking water', *Water and Environment Journal*, 11, 4, 257–63.
Hayes, N. 2002. 'Did manual workers want industrial welfare? Canteens, latrines and masculinity on British building sites, 1918–1970', *Journal of Social History*, 35, 3, 637–58.
Health Department 2013. Water Station Management Website, Kaohsiung City, http://203.65.117.243/index.asxp Accessed 03.07.2013.
Heller, M.A. 1998. 'The tragedy of the anticommons: property in the transition from Marx to markets', *Harvard Law Review*, 111, 3, 622–88.
Helm, D. and Rajah, N. 1994. 'Water Regulation: the periodic review', *Fiscal Studies*, 15, 2, 74–94.
Hennock, E.P. 2000. 'The urban sanitary movement in England and Germany, 1838–1914: a comparison', *Continuity and Change*, 15, 2, 269–96.
Heuts, F. and Mol, A. 2013. 'What is a good tomato? A case of valuing in practice', *Valuation Studies*, 1, 2, 125–46.

Heynon, N., Kaika, M., and Swyngedouw, E. (eds). 2005. *In the Nature of Cities: Urban political ecology and the politics of urban metabolism*. Routledge Press: London.

Heynon, N., McCarthy, J., Prudhom, S. and Robbins, P. 2007. *Neoliberal Environments: False Promises and Unnatural Consequences*. Routledge: London.

Holgate, J. 2005. 'Organizing migrant workers: a case study of working conditions and unionisation in a London sandwich factory', *Work Employment and Society*, 19, 3, 463–80.

Hosagrahar, J. 2010. 'Landscapes of water in Delhi: negotiating global norms and local cultures', in Sorensen, A. and Okarta, J. *Megacities: Urban Form, Governance and sustainability*. Springer: Japan, 111–132.

Huang, S-J., 2002. 'On Taiwan's Level of Urbanization', *Taiwanese Journal of Sociology*, 27, 163–205. (Text in Chinese).

INEGI. 2011a. Cuaderno Estadístico de la Zona Metropolitana del Valle de México 2011. Aguascalientes: INEGI.

INEGI. 2011b. Censo de Población y Vivienda 2010: Tabulados del Cuestionario Ampliado. Aguascalientes: INEGI.

INEGI. 2012. Directorio Estadístico Nacional de Unidades Económicas. Aguascalientes: INEGI. Retrieved from www3.inegi.org.mx/sistemas/mapa/denue/default.aspx Accessed 14.07.2013.

James, J. and Kerr, D. 2005. 'Prevention of childhood obesity by reducing soft drinks', *International Journal of Obesity*, 29, S54–S57.

Kaika, M. 2003. 'Constructing scarcity and sensationalising water politics: 170 days that shook Athens', *Antipode*, 35, 5, 919–54.

Kaika, M. and Swyngedouw, E. 2000. 'Fetishizing the modern city: the phantasmagoria of urban technological networks', *International Journal of Urban and Regional Research*, 24, 1, 120–38.

Kennett, D.J., Breitenbach, S.F., Aquino, V.V., Asmerom, Y., Awe, J., Baldini, J.U. and Haug, G.H. 2012. 'Development and disintegration of Maya Political systems in response to climate change', *Science*, 338, 788–91U.

Kjaernes, U., Harvey, M. and Warde, A. 2007. *Trust in Food: A Comparative and Institutional Analysis*. Palgrave: London.

Kloster, K. and Alba, F. de. 2007. 'El Agua en la Ciudad de México y el Factor de Fragmentación Política', *Perfiles Latinoamericanos*, 29, 137 59.

Kooy, M. and Bakker, K. 2008. 'Splintered networks: the contemporary and colonial waters of Jakarta', *Geoforum*, 39, 1843–58.

Kumar, N. and Aggarwal, S.C. 2003. 'Patterns of Consumption and Poverty in Delhi Slums', *Economic and Political Weekly*, 38, 50, 13–19.

Kurlansky, M. 1999. *Cod: A Biography of the Fish that Changed the World.* Vintage: London.

Lalwani, S., Dogra, T., Bhardwaj, D., Sharma, R. and Murty, O. 2006. 'Study on arsenic level in public water supply of Delhi using hydride generator accessory coupled with atomic absorption spectrophotometer', *Indian Journal of Clinical Biochemistry,* 21, 1, 70–76.

Latour, B. 1993. *The Pasteurisation of France*. Harvard University Press: London.

Latour, B. 2004. *The Politics of Nature: How to Bring the Sciences into Democracy*. Harvard University Press: Cambridge, MA.

Levasseur, J. and Maria, J. 2005. *Strategies for Water Supply in a Planned Urban Extension: A Case Study of the Dwarka Sub-city in Delhi*. Centre des Science Humaines: Delhi.

Li, W.D.H. 1998. *Housing in Taiwan: Agency and structure?,* Ashgate: Farnham.

Liao, C. 2007. 'Research on the development of the service equipment in the modern building of Taiwan during the Japanese colonial period', Master's Dissertation, Taiwan: Chung Yuan Christian University. (Text in Chinese). http://thesis.lib.cycu.edu.tw/ETD-db/ETD-search-c/view_etd?URN=etd-1226107-110555 Accessed 11.10.2013.

Lin, W. 1996. 'The factors of ethnic group fights in the Ching Dynasty', in Chang, Lee and Dai (eds) *The Selections of Taiwan's History*, Yu Shan Publisher: Taipei, 263–89. (Text in Chinese).
Liu, Y-H. 2004. *The History of Taiwan's Urban Development*. Feng Chia University: Taichung. (Text in Chinese). www.taichung.fcu.edu.tw/%A5x%C6W%B3%A3%A5%AB%B5o%AEi%A5v2004.pdf Accessed 14.10.2013.
Living Costs and Food Survey. 2009. Office for National Statistics. London.
Llorente, M. and Zerah, M.H. 2003. 'The Urban water sector: Formal versus informal suppliers in India', *Urban India,* 22, 1, 14–31.
Mahadevia, D., Liu, Z. and Yuan, X. 2012. 'Housing of urban migrants in India and China', *Environment and Urbanization Asia*, 3, 1, 143–63.
Mann, M. 2007. 'Delhi's Belly: on the management of water, sewage and excreta in a changing urban environment during the 19th century', *Studies in History*, 23, 1, 1–31.
Maria, A. 2006. 'The role of groundwater in Delhi's water supply: Interaction between formal and informal development of the water system, and possible scenarios of evolution', in J.H. Tellam (ed.) *Urban Groundwater Management and Sustainability*, Springer: Netherlands, 459–70.
Maria, A. 2008a. 'Urban water crisis in Delhi. Stakeholder responses and potential scenarios of evolution', *Idées pour le Débat*, 6, 1–23.
Maria, A. 2008b. 'Bricolage ou innovation? Etude de systèmes alternatifs d'approvisionnement en eau dans les quartiers périphériques de Delhi', *Idées pour le Débat*, 5, 1–32.
Marin, P. 2009. *Public-Private Partnerships for Urban Water Utilities: A Review of Experiences in Developing Countries.* World Bank: Washington DC.
Mazari-Hiriart, M., López-Vidal, Y., Ponce-de-León, S., Calva, J.J., Rojo-Callejas, F. and Castillo-Rojas, G. 2005. 'Longitudinal study of microbial diversity and seasonality in the Mexico City metropolitan area water supply system', *Applied and Environmental Microbiology*, 71, 9, 5129–37.
McCarthy, J. and Prudham, S. 2004. 'Neoliberal nature and the nature of neoliberalism', *Geoforum*, 35, 275–83.
McFarlane, C. 2012. 'Rethinking informality: politics, crisis and the city', *Planning Theory and Practice*, 13, 1, 89–108.
Mestre, E. 2001. 'The design of river basin organisations in Mexico – the example of Lerma-Chapala', paper to the *5th River Basin Management Workshop*. World Bank Institute: Washington DC.
Miele, M. and Evans, A. 2010. 'When foods become animals: Ruminations on ethics and responsibility in care-*full* practices of consumption', *Ethics, Place and Environment*, 13, 2, 171–90.
Millstone, E. and Russell, J. 1995. 'Lead toxicity and public health policy', *Journal of the Royal Society for the Promotion of Public Health*, 115, 6, 247–50.
Millward, R. 2000. 'The political economy of urban utilities in Britain 1840–1950', in M. Daunton (ed.) *Cambridge Urban History of Britain: Vol. III*, Cambridge University Press: Cambridge.
Millward, R. 2005. *Private and Public Enterprise in Europe. Energy, Telecommunications and Transport*. Cambridge University Press: Cambridge.
Millward, R. 2007. 'Urban water supplies c 1820–1950: the dilemma of private companies', *Histoire, Economie, Société*, 2, 111–28.
Ministry of Interior, 2011. *The Statistical Yearbook, 2011*. http://sowf.moi.gov.tw/stat/month/m1-07.xls Accessed 06.07.2013.
Ministry of Economy 2013. *Economic statistics*, https://2k3dmz2.moea.gov.tw/gwWeb/ Accessed 05.07.2013.

Mintz, S. 1985. *Sweetness and Power: The Place of Sugar in Modern History*. Penguin: London.

Molle, F., Mollinga, P.P. and Wester, P. 2009. 'Hydraulic bureaucracies and the hydraulic mission: flows of water, flows of power', *Water Alternatives*, 2, 3, 328–49.

O'Connor, J. 1998. *Natural Causes: Essays in Ecological Marxism*. Guilford Press: New York.

Ostrom, E., Dietz, T.E., Dolšak, N.E., Stern, P.C., Stonich, S.E. and Weber, E.U. 2002. *The Drama of the Commons*. National Academy of the Sciences: Washington DC.

Parker, D.J. and Derrick Sewell, W.R. 1988. 'Evolving water institutions in England and Wales: an assessment of two decades of experience', *Natural Resources Journal*, 28, 751–85.

Peck, J. 2013. 'For Polanyian economic geographies', *Environment and Planning A*, 45, 1545–68.

Peña, F. 2004. Pueblos Indígenas y Manejo de Recursos Hídricos en México. *Revista Mad, 11*, 1–10.

Peña, C.L.C. de la. 2011. 'Crisis del modelo de gestión del agua del Sistema Cutzamala: derecho humano al agua, disponibilidad y conflictos', in E.G. Gómez, N.B.G. Ramírez and S.V. Velázquez (eds) *Gestión Social y Procesos Productivos* (pp. 87–110). Universidad Autónoma del Estado de Morelos: Cuernavaca.

Peng, L. 2013. 'Effects of anti-dam campaigns on institutional capacity: a case study of Meinung from Taiwan', *Paddy and Water Environment*, 11, 1–4, 353–67.

Planning Department 2009. Economic Survey of Delhi. Government of NCT of Delhi: Delhi.

Planning Department 2013. Economic Survey of Delhi. Government of NCT of Delhi: Delhi.

Polanyi K. 1944. *The Great Transformation: The Political and Economic Origins of Our Time*. Beacon Books: New York.

Polanyi, K. 1957. 'Aristotle discovers the economy', in Polanyi, K., Arensberg, C.M. and Pearson, H.W. (eds) *Trade and Market in the Early Empires*. The Free Press; New York, 64–94.

Polanyi, K. 1966. *Dahomey and the Slave Trade: An Analysis of an Archaic Economy*. American Ethnological Society Monograph, 42. University of Washington Press.

Pomeranz, K. 2000. *The Great Divergence: China, Europe, and the Making of the Modern World Economy*. Princeton University Press: Princeton.

Ponte, S. and Gibbon, P. 2005. 'Quality standards, conventions and the governance of global value chains', *Economy and Society*, 34, 1, 1–31.

Prashad, V. 2001. 'The technology of sanitation in colonial Delhi', *Modern Asian Studies*, 35, 1, 113–55.

Raghupathi, U. 2003. 'Small private water providers: An alternative solution for the poor' *Shelter*, 6, 3, 23–7.

Read, C.J. 2012. 'A place in the city: Narratives of 'emplacement' in a Delhi resettlement neighbourhood', *Ethnography*, 13, 1, 87–101.

Right2Water. 2015. www.right2water.ie/ Accessed 07.05.2014.

Rong, F. 2011. 'Governing water in China: implications from four case studies', *International Law and Regulation Working Paper*, 4, August. University of California, San Diego.

Rose, C. 1991. *The Dirty Man of Europe: The Great British Pollution Scandal*. Simon & Schuster: London.

Roy, A. 2005. 'Urban informality: toward an epistemology of planning', *Journal of the American Planning Association*, 71, 2, 147–58.

Roy, A. 2008. 'Why India cannot plan its cities: informality, insurgence and the idiom of urbanisation', *Planning Theory*, 8, 1, 76–87.

Roy, D. 2013. 'Negotiating marginalities: right to water in Delhi', *Urban Water Journal*, 10, 2, 97–104.

Royte, E. 2011. *Bottlemania: Big Business, Local Springs, and the Battle over America's Drinking Water*. Bloomsbury Publishing: New York.

Ruddan, B. 1985. *The New River: A Legal History*. Clarendon Press: Oxford.

SACM, 2012. *Empresas Concesionarias*. www.sacmex.df.gob.mx:8080/web/sacm/empresas_concesionarias Accessed 02.07.2012.

Salzman, J. 2014. *Drinking Water: A History*. Gerald Duckworth: London.

Schatzki, T. 1996. *Social Practices: A Wittgensteinian Approach to Human Activity and the Society*. Cambridge University Press: Cambridge.

Schatzki, T. 2005. 'Peripheral vision: the sites of organization', *Organization Studies*, 26, 3, 465–84.

Schwartz, K., Luque, M.T., Rusca, M. and Ahlers, R. 2014. '(In)formality: the meshwork of water provisioning', *WIREs Water*, doi:10.1002/wat2.1056

Schwartz, P. 1966. 'John Stuart Mill and laissez faire: London water', *Economica*, February, 71–83.

Shaban, A. 2008. *Water Poverty in Urban India: A Study of Major Cities*. Paper presented at the UGC-Summer Programme.

Sharan, A. 2011. 'From source to sink: "Official" and "improved" water in Delhi, 1868–1956', *Indian Economic Social History Review*, 48, 425–62.

Sharma, A. 2014. *Standards, Technology and Public Participation: Regulation Making for Bottled Water Quality in India*. PhD Thesis/ Jawaharlal Nehru University, New Delhi.

Sharma, A. and Bhaduri, S. 2012. 'Choice or necessity: the dynamics of bottled water consumption in urban India', Paper to the *International Symposium on Anthropology of Global Issues*, New Delhi, 1–3 April.

Sharma, Y. 2009. 'The circuit of life: water and water reservoirs in pre-modern India', *Studies in History*, 25, 1, 69–108.

Sheail, J. 1986. 'Government and the perception of reservoir development in Britain: a historical perspective', *Planning Perspectives*, 1, 1, 45–60.

Shove, E. 2003. *Comfort, Cleanliness and Convenience: The Social Organisation of Normality*, Berg: Oxford.

Sivam, A. 2003. 'Housing supply in Delhi', *Cities*, 20, 2, 135–41.

Sosa, F. 2010a. 'Impacts of water-management decisions on the survival of a city: from ancient Tenochtitlan to modern Mexico City', *International Journal of Water Resources Development*, 26, 4, 675–87.

Sosa, F. 2010b. 'Exploring the risks of ineffective water supply and sewage disposal: A case study of Mexico City', *Environmental Hazards*, 9, 2, 135–46.

Sosa, F. 2012. 'Assessing water quality in the developing world: an index for Mexico City, in Voudouris, K. and Voutsa, D. (eds) *Water Quality Monitoring and Assessment*, 495–508.

Stern, N. 2014. *Better Growth, Better Climate: The New Climate Economy Report*, World Resources Institute, Washington DC, www.newclimateeconomy.report Accessed 02.10.2014.

Sultana, F. and Loftus, A. 2011. *The Right to Water: Politics, Governance and Social Struggles*. Routledge: London.

Sunderland, D. 1999. '"A monument to defective administration"?: The London Commission of Sewers in the Early 19th century', *Urban History*, 26, 3, 349–72.

Swiss Gas and Water Association, 2006. *Comparison of the Environmental Impact of Drinking Water vs. Bottled Mineral Water*.

Swyngedouw, E. 2004. *Social Power and the Urbanization of Water: Flows of Power*. Oxford University Press: Oxford.

Taiwan Provincial Council. 1962. Taiwan Provincial Council Records, vol. 8, no. 13, p.343, http://catalog.digitalarchives.tw/item/00/17/be/2c.html Accessed 02.10.2014.

Taiwan Provincial Government. 1964. *The Plan for Public Wells and Toilet Building*, The Health Department, Taiwan.

Taiwan Water Association. 2006a. *Taiwan's Tap Water*, Taiwan Water Association: Taipei. (Text in Chinese).

Taiwan Water Association, 2006b. *The Research Goal and Strategy of the 21st Century's Water Supply Development*. Water Resource Agency, Ministry of Economic Affairs: Taiwan. (Text in Chinese).

Taiwan Water Corporation. 2007. *Environmental Sustainability Resource Efficiency: Report on Tai Water Reasonable Water Rate Programs*. (Text in Chinese). www9.water.gov.tw/ch/files/waterch/水價合理調整_03.doc Accessed 14.11.2013.

Taiwan Water Corporation. 2011. *Board Meeting Report numbered 15(4)*, www.water.gov.tw/torule/file/1504Record.pdf. Accessed 14.11.2013.

Taiwan Water Corporation 2013. *Special Report on Water Quality Management*, www7.water.gov.tw/book01/01_08_01_01.htm Accessed 02.07.2013.

Taylor Nelson Sofres. 2010. *Verband Deutscher Mineralbrunnen e.V – Mineralwasser 2010*, www.vdm-bonn.de/home.html Accessed 02.02.2015.

Taylor, V. and Trentmann, F. 2011. 'Liquid politics: water and the politics of everyday life in a modern city', *Past and Present*, 211, 1, 199–241.

Thrift, N. 1996. *Spatial Formations*. (Theory, Culture and Society Special Edition). Sage: London.

Tortajada, C. 2003. 'Water management for a megacity: Mexico City Metropolitan Area', *Ambio*, 32, 2, 124–29.

Tortajada, C. 2006. 'Who has access to water? A case study of the Mexico City Metropolitan Area', *Human Development Report, Occasional Paper*, UNDP: New York.

Toxics Link. 2014. *A Report on the Toxic Load in the Yamuna River in Delhi*, http://toxicslink.org/docs/Yamuna-Report-Toxics-Link_12-12-2014.pdf Accessed 24.02.2015.

Trentmann, F. and Taylor, V. 2005. 'From users to consumers – water politics in nineteenth century London', in Trentmann, F. (ed.) *The Making of the Consumer: Knowledge, Power and Identity in the Modern World.* Berg Publishers: Oxford, 53–79.

UNAM. 2010. Evaluación externa del diseño e implementación de la política de acceso al agua potable del Gobierno del Distrito Federal. Mexico DF.

United States Government. 1902. *Public Heath Reports*, Washington, DC., www.ncbi.nlm.nih.gov/pmc/articles/PMC1998952/pdf/pubhealthreporig04082-0001.pdf Accessed 21.11.2014.

Urry, J. 2010. 'Consuming the planet to excess', *Theory, Culture & Society*, 27, 2–3, 191–212.

Urry, J. 2011. *Climate Change and Society*. Polity: Cambridge.

Urry, J. 2014. 'The Problem of Energy', *Theory, Culture & Society*, 31, 5, 3–20.

Vargas, S. 2008. El conflicto y la negociación en la percepción de los usuarios del agua en la cuenca Lerma-Chapala, 1999–2004. *Revista Legislativa de Ciencias Sociales y de Opinión Pública*, 1, 1, 155–84.

Verband Deutscher Mineralbrunnern. 2011. *Facts and Figures of the German Mineral Water and Soft Drink Industry 2011*, Powerpoint presentation, received via personal correspondence, www.vdm-bonn.de/home.html

Walker, A. 2009. *The Independent Review of Charging for Household Water and Sewage Services: Final Report*. OFWAT/DEFRA, www.defra.gov.uk/environment/quality/water/industry/walkerreview/index Accessed 02.10.2014.

Wang, J.-H. and Huang, S. 2013. 'State re-scaling and water governance: the South-North Water Transfer Project in China', Paper to the International Workshop, *Properties of Water: Qualities, Rights Over Water Resources, Sustainability*, University of Essex, October 3–4.

Wang, S. 2001. *The History of Water Resources in Taiwan Area*, Taiwan Sheng wen xian wei yuan hui. Taiwan. (Text in Chinese).

Warde, A. 2005. 'Consumption practices and theories of practice', *Journal of Consumer Culture*, 5, 2, 131–53.

Water Resource Agency. 2009. *Water Consumption Statistical Database*, Ministry of Economic Affairs, http://wuss.wra.gov.tw/annmisc.aspx Accessed 09.02.2014.

Water Resource Agency. 2010. *Annual Report 2010*, http://wuss.wra.gov.tw/annualreports/ Accessed 09.02.2014.

Water Resource Agency. 2013. *Annual Report 2013*, http://wuss.wra.gov.tw/annmisc.aspx Accessed 09.02.2014.

Whitley, R. 1999. *Divergent Capitalisms: The Social Structure and Change of Business Systems*. Oxford University Press: Oxford.

Wilk, R. 2006. 'Bottled water: the pure commodity in the age of branding', *Journal of Consumer Culture*, 6, 3, 303–25.

Wittfogel, K.A. 1963. *Oriental Despotism: A Comparative Study of Total Power*. Yale University Press.

Wu, C.-Y. 1997. 'A study on drinking water of Kaohsiung region', Masters Dissertation, Geography Department, National Taiwan Normal University, Taipei.

Zheng, C., Liu, J., Cao, G., Kendy, E., Wang, H. and Jia, Y. 2010. 'Can China cope with its water crisis? Perspectives from the North China Plain', *Ground Water*, 48, 3, 350–54.

Zerah, M.H. 1998. 'How to assess the quality dimension of urban infrastructure: the case of water supply in Delhi', *Cities*, 15, 4, 285–90.

Zerah, M.H. 2000. *Water: Unreliable Supply in Delhi*. Manohar: Delhi.

Index

For Product Safety Concerns and Information please contact our EU representative GPSR@taylorandfrancis.com
Taylor & Francis Verlag GmbH, Kaufingerstraße 24, 80331 München, Germany

www.ingramcontent.com/pod-product-compliance
Ingram Content Group UK Ltd.
Pitfield, Milton Keynes, MK11 3LW, UK
UKHW022125190926
475702UK00013B/1652
* 9 7 8 1 1 3 8 3 0 4 9 7 0 *